P9-CBG-214

The Digestive System

Other titles in
Human Body Systems

The Circulatory System by Leslie Mertz

The Endocrine System by Stephanie Watson and Kelli Miller

The Lymphatic System by Julie McDowell and Michael Windelspecht

The Muscular System by Amy Adams

The Nervous System and Sense Organs by Julie McDowell

The Reproductive System by Kathryn H. Hollen

The Respiratory System by David Petechuk

The Skeletal System by Evelyn Kelly

The Urinary System by Stephanie Watson

OCLS/MULLICA HILL BRANCH
389 WOLFERT STATION ROAD
MULLICA HILL, NJ 08062

The Digestive System

Michael Windelspecht

HUMAN BODY SYSTEMS
Michael Windelspecht, Series Editor

Greenwood Press
Westport, Connecticut • London

Library of Congress Cataloging-in-Publication Data

Windelspecht, Michael, 1963–
The digestive system / Michael Windelspecht.
p. cm.—(Human body systems)
Includes bibliographical references and index.
ISBN 0–313–32680–0 (alk. paper)
1. Digestive organs—Diseases. 2. Digestive organs. 3. Nutrition.
I. Title. II. Human body systems.
RC801.W557 2004
612.3—dc22 2004040446

British Library Cataloguing in Publication Data is available.

Copyright © 2004 by Michael Windelspecht

All rights reserved. No portion of this book may be reproduced, by any process or technique, without the express written consent of the publisher.

Library of Congress Catalog Card Number: 2004040446
ISBN: 0–313–32680–0

First published in 2004

Greenwood Press, 88 Post Road West, Westport, CT 06881
An imprint of Greenwood Publishing Group, Inc.
www.greenwood.com

Printed in the United States of America

The paper used in this book complies with the Permanent Paper Standard issued by the National Information Standards Organization (Z39.48–1984).

10 9 8 7 6 5 4 3 2 1

Illustrations, unless otherwise credited, are by Sandy Windelspecht.

The *Human Body Systems* series is a reference, not a medical or diagnostic manual. No portion of this series is intended to supplement or substitute medical attention and advice. Readers are advised to consult a physician before making decisions related to their diagnosis or treatment.

To my wife, Sandy,
For all her love and support

Contents

Series Foreword

Human Body Systems is a ten-volume series that explores the physiology, history and diseases of the major organ systems of humans. An organ system is defined as a group of organs that physiologically function together to conduct an activity for the body. In this series we identify ten major functions. These are listed in Table F.1, along with the name of the organ system responsible for the activity. It is sometimes difficult to specifically define an organ system, because many of our organs have dual functions. For example, the liver interacts with both circulatory and digestive systems, the hypothalamus acts as a junction between the nervous and endocrine systems, and the pancreas has both digestive and endocrine secretions. This complex interaction of organs and tissues in the human body is still not completely understood.

This series is unique in that it provides a one-stop reference source for anyone with an interest in the human body. Whereas other references frequently cover one aspect of human biology, from anatomy and physiology to the prevention of diseases, this series takes a more holistic approach. Each volume not only includes a physiological description of how the system works from the cellular level upward, but also a historical summary of how research on the system has changed since the time of the ancients. This is an important aspect of the series, and one that is frequently overlooked in modern textbooks. In order to understand the successes and problems of modern medicine, it is first important to recognize not only the achievements of the past but also the misunderstandings and challenges of the pioneers in medical research.

For example, a visit to any major educational institution reveals large lec-

TABLE F.1. Organ Systems of the Human Body

Organ System	General Function	Examples
Circulatory	Movement of chemicals through the body	Heart
Digestive	Supply of nutrients to the body	Stomach, small intestine
Endocrine	Maintenance of internal environmental conditions	Thyroid
Lymphatic	Immune system, transport, return of fluids	Spleen
Muscular	Movement	Cardiac muscle, skeletal muscle
Nervous	Processing of incoming stimuli and coordination of activity	Brain, spinal cord
Reproductive	Production of offspring	Testes, ovaries
Respiratory	Gas exchange	Lungs
Skeletal	Support, storage of nutrients	Bones, ligaments
Urinary	Removal of waste products	Bladder, kidneys

ture halls, where science instructors present material to the students on the anatomy and physiology of the human body. Sometimes these classes include laboratory sessions, but in the study of human biology, especially for students who are not bound for professional schools in medicine, the student's exposure to human biology typically centers on a two-dimensional graphic. Most educators accept this process as a necessary evil of the educational system, but few recognize that, in fact, the large lecture classroom is the product of a change in Egyptian religious beliefs before the start of the current era. During the decline of the Egyptian empire and the simultaneous rise of the ancient Greek culture, the Egyptian religious organizations began to forbid the dissection of the human body. This had a twofold influence on medicine. First, the ending of human dissections meant that medical professionals required lectures from educators, instead of participation in laboratory-based education, which led to the birth of the lecture hall. The second consequence would plague modern medicine for a thousand years. Stripped of their access to human cadavers, researchers studied other "lesser" animals and extrapolated their findings to humans. The practices of the ancient Greeks were passed on over the ages and became the basis for the study of modern medicine. These traditions continue to this day throughout the educational institutions of the world.

The history of human biology parallels the development of modern science. In the seventeenth century, William Harvey's study of blood circulation challenged the long-standing belief of the ancient Greeks that blood was produced in the liver and consumed in the tissues of the body. Harvey's pioneering experimental work had a strong influence on others, and within a century the legacy of the ancient Greeks had collapsed. In the eighteenth century, a group of chemists who focused on the chemical reactions of the human body, called the iatrochemists, began to apply chemical laws to human physiology. They were joined by the iatrophysicists, who believed that the human body must operate under the physical laws of the universe. This in turn led to the beginnings of organic chemistry and biochemistry in the nineteenth century, as scientists focused on identifying the building blocks of living cells and the chemical reactions that they utilize in their metabolism.

In the past century, especially in the last three decades, the rapid advances in technology and scientific discovery have tended to separate most sciences from the general public. Yet despite an ongoing trend to leave the majority of the physical sciences to the scientists, interest in human biology has actually increased among the general population. This is primarily due to medical discoveries that increase not only lifespan but also healthspan, or the number of years that people live disease free. But another important aspect of this trend is the desire among the general public to be able to ask intelligent questions of their physicians and seek additional information on prescribed medications or procedures. In many cases, this information serves as a system of checks and balances on the medical profession, ensuring that the patient is kept well informed and aware of the fundamentals regarding the procedure.

This is one of the most remarkable ages in the study of human biology. The recently announced completion of the Human Genome Project is an indication of how far biology has progressed. Barely fifty years ago, scientists were first discovering the structure of DNA. They now are in possession of an entire encyclopedia of human genetic information, and although they are not yet exactly sure what the content reveals, scarcely a week goes by without a researcher announcing a medical discovery that was made possible by the availability of the complete human genetic sequence. Coupled to this are the advances in the development of pharmaceuticals and treatments that were unheard of less than a decade ago.

But these benefits to society do not come without a cost. The terms stem cells, cloning, and gene therapy no longer belong to the realm of science fiction. They represent advances in the sciences that may hold the key to increased longevity. However, in many cases they also produce ethical and moral questions of society: Where do medical researchers obtain the embryonic stem cells for their work? Who will determine if humans can be cloned? What are the risks of transgenic organisms produced by gene therapy? These are just a few of the potential conflicts that face modern society. Only a well-educated general public can intelligently survey the pros and cons of an ethical or moral decision regarding medical science. Armed with information, concerned people can participate in the democratic process of informing their elected officials of their concerns. Science education is an important aspect of citizenship, and thus the need for series such as this to present information to the general public.

This volume covers the biology of the digestive system, which represents one of the largest organ systems in the human body. It is responsible for the processing of ingested food and liquids. The cells of the human body all require a wide array of chemicals to support their metabolic activities, from organic nutrients used as fuel to the water that sustains life at the cellular level. The digestive system not only chemically reduces the compounds in food into their fundamental building blocks effectively, but also acts to retain water and excrete undigested material. The efficiency of these metabolic processes varies not only between individuals, but also changes based on aging, disease, environmental factors, and conditions such as pregnancy. At some point in their lives, all people will suffer from an ailment of the digestive system. This volume provides useful reference information on all aspects of the digestive processes.

The ten volumes of *Human Body Systems* are written by professional authors who specialize in the presentation of complex scientific ideas to the general public. Although any book on the human body must include the terminology and jargons of the profession, the authors of this series keep it to a minimum and strive to explain the concepts clearly and concisely. The series is ideal for the public libraries, as well as for secondary school and introductory college libraries. In addition, medical professionals or anyone with an interest in human biology would find this series a useful addition to their personal library.

Michael Windelspecht
Blowing Rock, North Carolina

Acknowledgments

I would first like to acknowledge the editorial and production staff at Greenwood Publishing for their invaluable assistance in the preparation of this series, and specifically this volume. Special thanks to my acquisitions editor, Debby Adams, for her patience with my endless series of questions. I would also like to thank Liz Kincaid, the photo researcher for this series, for her hard work obtaining the images contained in this volume.

I owe a special note of appreciation to my wife, Sandy. Not only does she support me in my writing projects, but she also serves as my illustrator. Her patience is borderline divine, especially because I have a terrible habit of changing graphics in midstream.

I would also like to thank the other authors of this series for their assistance and patience with the process. The finished product is far better than what I had initially envisioned when I began the series, and this is due to their dedication and abilities as science writers.

As always I would like to thank Greenwood Publishing for the opportunity to contribute to this series. An understanding of science and medicine is one of my passions, and I am very content that publishers such as Greenwood recognize the need to provide quality educational materials for a general audience.

Introduction

One of the unifying characteristics of all living organisms is their ability to process nutrients from the environment into the chemical compounds found within the cells. This processing of nutrients is commonly called metabolism. Plants, animals, fungi, and bacteria have all evolved different strategies for supplying the energy and chemical needs of the organism. Simply stated, digestion is the breakdown of food particles into their fundamental building blocks. As heterotrophic organisms, or those that rely on others as a source of energy, animals have evolved a wide range of digestive systems to accommodate their environmental needs. The internal body plan of an animal species is frequently defined by its digestive system. The purpose of this volume is to examine the structure and function of the digestive systems in the species *Homo sapiens*, or humans.

The human digestive system is a highly evolved group of organs that interact to provide all of the approximately 63 trillion cells of the human body with the nutrients for growth, reproduction, and metabolism of compounds needed for daily activity by the body. Despite the obvious importance of the digestive system, until the past century, scientists were slow in researching the physiology of the gastrointestinal tract. For most of recorded history, religious beliefs and customs inhibited the ability of scientists to examine the body after death. Without models to study, many misconceptions were developed on the role of the digestive organs, a number of which persisted until the seventeenth and eighteenth centuries. However, with the advent of technology and biotechnology, and their tremendous influence on twentieth- and twenty-first-century medicine, new diagnostic tools and medicines are available to researchers. The result is a vastly accelerated pace of discovery. Over the past one hundred years, medical professionals and scientists have made up for lost time and have made tremendous strides in understanding the physiology of the digestive system. Physicians can now treat the causes of gastrointestinal diseases, and not simply the symptoms of the disease, as had been the case for much of human history. In fact, on any given day, it would be difficult not to find an article in the paper or major news magazine that did not relate to nutrition or a human disease associated with the digestive tract. Thus the need for a reference book that provides information to the general audience on the workings of the digestive system.

The digestive system is not a stand-alone system. The primary digestive organs—the mouth, esophagus, stomach, and small and large intestine—not only interact in some degree with each other, but also receive signals from other organs of the body. The accessory organs—the liver, pancreas, salivary glands and gall bladder—supply chemicals necessary for the nutrient processing. The liver and pancreas are active with other systems of the body as well, such as the endocrine and circulatory systems. The digestive system is partially under the control of the nervous system, but also is influenced by the hormones secreted by the endocrine system. Because nutrients absorbed by the digestive system must be transported throughout the body, the gastrointestinal tract interacts with the circulatory and lymphatic system for the transport of water-soluble and fat-soluble nutrients, respectively. Finally, the urinary system removes some of the waste products of nutrient metabolism by the liver.

This work is not solely a discussion of the anatomy and physiology of the digestive organs. Due to the digestive system's central role in nutrient processing, a significant amount of the material contained in this volume addresses the role of nutrition in human health, plus the history of discovery for select nutrients. Oddly enough, scientists have only recently recognized that there was a distinct link between nutrient deficiencies and disease. This is especially the case for the vitamins and minerals, whose relationship to disease only became known in the past 100 years. This is an ongoing area of research, with thousands of scientists at numerous research facilities investigating the biochemistry and genetics of nutrition. The first chapter of this volume provides an overview of nutrition.

This volume is part of a ten-volume series on the human body. It is designed as a reference volume for anyone interested in obtaining an overview of the physiology, history and ailments of the digestive system. Following a list of interesting facts about the digestive system, the volume is organized into three primary sections. The first section (Chapters 1–4) examines the basic anatomy and physiology of the digestive system. In the second section (Chapters 5–7), a discussion of the history of discovery of nutrition and the digestive system from the ancients to the present is covered. In the third section (Chapters 8–12), ailments and diseases of the digestive tract, including the accessory organs, are discussed. At the end of the volume is a list of commonly used acronyms, followed by a glossary of important terms, a list of organizations and Web sites, and a bibliography. An index is also provided at the end to facilitate cross-referencing of topics.

This work is targeted at the general science audience and as such an attempt has been made to describe as many of the medical and scientific concepts in common language. The glossary at the end of the work provides definitions or examples of key terms. **Bold** type indicates the first use of the term in the volume. The organization of this volume and series makes this work attractive for secondary school libraries, undergraduate higher education colleges, and universities where students may be seeking general information on the digestive system. In addition, community libraries that wish to possess a general reference volume on the digestive system, as well as anyone with an interest in science, history, or medicine, will find this work a useful addition to their collection.

INTERESTING FACTS

- The salivary glands can produce up to 1.5 quarts (1,500 milliliters) of saliva daily.
- Food remains in the esophagus for as little as 5 seconds before entering the stomach.
- The human stomach can hold as much as 2.1 quarts (2 liters) of food.

- The stomach produces 2.12 quarts (2 liters) of gastric juice daily.
- Gastric juice is 100,000 times more acidic than water and has about the same acidity as battery acid.
- The small intestine in an adult can reach 3.28 yards (3 meters) in length, while the large intestine is only about 1.64 yards (1.5 meters) long.
- Every square millimeter of the small intestine can contain 40 villi and 200 million microvilli.
- Food remains in the small intestine for 3–5 hours on average, during which most nutrients are removed.
- The intestines receive over 10 quarts (9 liters) of water daily, of which almost 95 percent is recycled back into the body.
- As little as 10–15 percent of the iron we eat in food is absorbed into the human body.
- The color of the feces and urine is actually the result of the breakdown of red blood cells in the liver.
- The human liver may weigh up to 3 pounds (1.4 kilograms) and may regenerate up to 60 percent of its mass in case of injury.
- The herpes simplex virus is present in 80 percent of the U.S. population.

- In 1993, 403,000 residents were infected by a single outbreak of *Cryptosporidium*.
- An adult roundworm can lay 200,000 eggs a day in the lumen of the intestine.
- Some species of adult tapeworms can reach lengths of 11 yards (10 meters).

Digestion at the Cellular and Molecular Level

When the term digestion is mentioned, it is natural to think about the action of the mouth, stomach and small intestine in the processing of food for energy. While the action of these organs is without doubt important in the breakdown of food, they actually are the result of complex processing mechanisms at the cellular level. The next three chapters will examine the physiology of the digestive system at the organ level. However, to effectively understand the structure and function of the digestive system it is necessary to first understand the cellular and molecular basis of nutrient processing. This chapter will establish the fundamental principles of nutrient **metabolism** at the cellular level. These principles will continue as an underlying theme throughout the remainder of the volume.

The purpose of digestion is to process food by breaking the chemical bonds that hold the nutrients together. This is necessary so that the body has an adequate source of energy for daily activity, as well as materials for the construction of new cells and tissues. Since these nutrients arrive in the digestive system as the tissues of previously living organisms, they are rarely in the precise molecular structure needed by a human body. For example, the blood of cows and chickens has evolved over time to meet the precise metabolic needs of the organism. When the tissues of these animals are consumed, our bodies must chemically alter the proteins and other nutrients found in the animal's blood to form human blood proteins such as **hemoglobin**. As is the case with almost all the nutrients (with the exception of water, minerals, and some vitamins), the body breaks down the nutrient into its fundamental building blocks, transports the digested nutrient into circulatory and lymphatic systems, and eventually uses these nutrients in the cells of the body for either energy or metabolic processes. Before proceeding into a discussion into how these reactions occur, it is first necessary to discuss the basic nutrient classes. The following information is designed to provide an overview of the major nutrient classes. Throughout the volume additional information will be provided with regard to the operation of each organ and during the discussions of disease states. Readers who wish for more detailed information on human nutrition should consult the Organizations and Web Sites chapter at the back of this volume, or the bibliography, for additional sources of information.

CLASSES OF NUTRIENTS

There are six general classes of nutrients: carbohydrates, fats, proteins, water, vitamins and minerals. Carbohydrates, fats and proteins are characterized as energy nutrients. These **organic** (or carbon-containing) molecules are responsible for providing our bodies with the majority of the energy needed for daily metabolic reactions. This does not mean that the remaining nutrient classes are not important in energy reactions within the body. In fact many of these, such as some of the B vitamins and water, are crucial to the efficient operation of the energy pathways. However, our bodies do not get energy from these nutrients directly.

Carbohydrates, proteins, and fats all contain energy in the carbon-carbon bonds of their molecules. The energy of these bonds is measured in a unit of heat measurement called the calorie. A calorie is the amount of energy required to raise 1 gram of pure water by 1 degree Celsius at sea level. However, this is a relatively small unit of measurement and thus for nutritional analysis the term **kilocalorie** (1,000 calories) is frequently used. When one examines the ingredient label of a prepared food, such as a soft drink, the listed calorie value is actually in kilocalories, also called *kcals*. Organic molecules contain a large number of carbon-carbon bonds, and are therefore an excellent source of metabolic energy.

ENERGY FROM ENERGY NUTRIENTS

Cells have a variety of mechanisms for releasing the energy contained within the carbon-carbon bonds of organic molecules. Some cells are **anaerobic** and can obtain small amounts of energy without the assistance of oxygen. However, the majority of the cells of the body utilize a complex metabolic pathway called aerobic respiration. Aerobic respiration consists of three main series of reactions: **glycolysis**, the Krebs cycle, and the electron transport chain (ETC). The Krebs cycle and ETC occur in the **mitochondria** of the cell and use oxygen to regenerate a cellular energy molecule

The Lysosome

In some cells there is a specialized **organelle** called the lysosome that is involved in the digestive process. In simple animal cells, such as amoebas, the lysosome performs many of the same functions as the organs of the digestive system of higher animals, meaning that it breaks down incoming nutrients into their fundamental building blocks. In animals such as humans, the lysosome does not need to perform this function since the function of the gastrointestinal tract is to supply the cells of the body with digested nutrients. Instead, this organelle performs a unique form of cellular digestion in that it digests, or recycles, worn out or damaged cellular components such as mitochondria.

One can imagine that this type of activity must be highly regulated by the cell. In fact there is a number of disease states associated with abnormal lysosome activity. One of these is Tay-Sachs disease, a degenerative disorder of the nervous system. This disease causes a lysosome to fill with a form of lipid that makes the lysosome burst, releasing the digestive enzymes into the cytoplasm of the cell, thus killing it.

called *adenosine triphosphate* (ATP). In the Krebs cycle, the carbon-carbon bonds are broken and a small amount of ATP is generated. The remaining carbon is combined with oxygen to form carbon dioxide, a waste product. The details of aerobic respiration will be covered in greater detail in the respiratory volume of this series.

The aerobic respiration pathways are capable of utilizing most organic molecules as an energy source. However, in order for the nutrients to enter the pathway they must first be broken down into their fundamental building blocks (see "The Lysosome"). Proteins and carbohydrates are actually long repetitive chains of individual building blocks called *monomers*. These monomers are linked by chemical bonds into long polymers (see Table 1.1),

TABLE 1.1. The Energy Nutrients

Energy Nutrient	Monomer	Polymer	General Enzyme	Energy per gram
Carbohydrates	Monosaccharides	Polysaccharides	amylases	4 kcal
Proteins	Amino acids	Proteins	proteases	4 kcal
Fats and Lipids	N/A	N/A	lipases	9 kcal

This table lists some important facts regarding the energy nutrients. The N/A under fats and lipids reflects the fact that these molecules do not form complex structures in the same manner as carbohydrates and proteins. The energy per gram is an approximation and varies depending on the metabolic properties of the cell and individual.

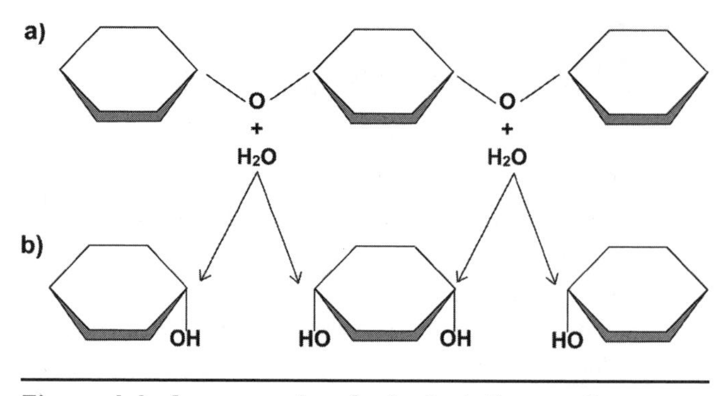

Figure 1.1. An example of a hydrolytic reaction.
In this reaction water is being added to the polysaccharide (a) forming three monosaccharides (b).

which first must be broken down. Water is used to break the bonds linking the monomers in a process called **hydrolysis**. Figure 1.1 demonstrates the hydrolysis of a starch molecule into individual glucose units. Cells use the reverse of this process, called **dehydration synthesis** or condensation reactions, to form more complex molecules from the monomers.

Carbohydrates

In the study of nutrition, carbohydrates are frequently abbreviated as CHO, which reflects the fact that this nutrient class contains the elements carbon, hydrogen, and oxygen. All carbohydrates possess carbon, hydrogen, and oxygen in a 1:2:1 ratio, respectively. For example, the molecular formula for glucose is $C_6H_{12}O_6$. Carbohydrates are the short-term energy molecules of the human body and are the preferred fuel of the aerobic respiration pathways.

All carbohydrates are made up of one of three building blocks, or *monosaccharides.* The most common of these is glucose, with the other two being fructose (the monosaccharide associated with the sweet taste) and galactose (sometimes called a milk sugar). The structure of glucose is shown in Figure 1.2. Glucose is the preferred energy molecule for aerobic respiration, and later chapters will illustrate how well-adapted the human digestive system is in extracting this nutrient from foods and delivering it to the cells. All of the monosaccharides are water-soluble and easily transported by the circulatory system.

Monosaccharides are linked together in pairs by chemical bonds to form the disaccharides. All *disaccharides* contain at least one glucose unit in their structure. There are three different disaccharides: maltose (glucose-glucose), sucrose (glucose-fructose) and lactose (glucose-galactose). An ailment of the digestive system called lactose intolerance will be discussed in greater

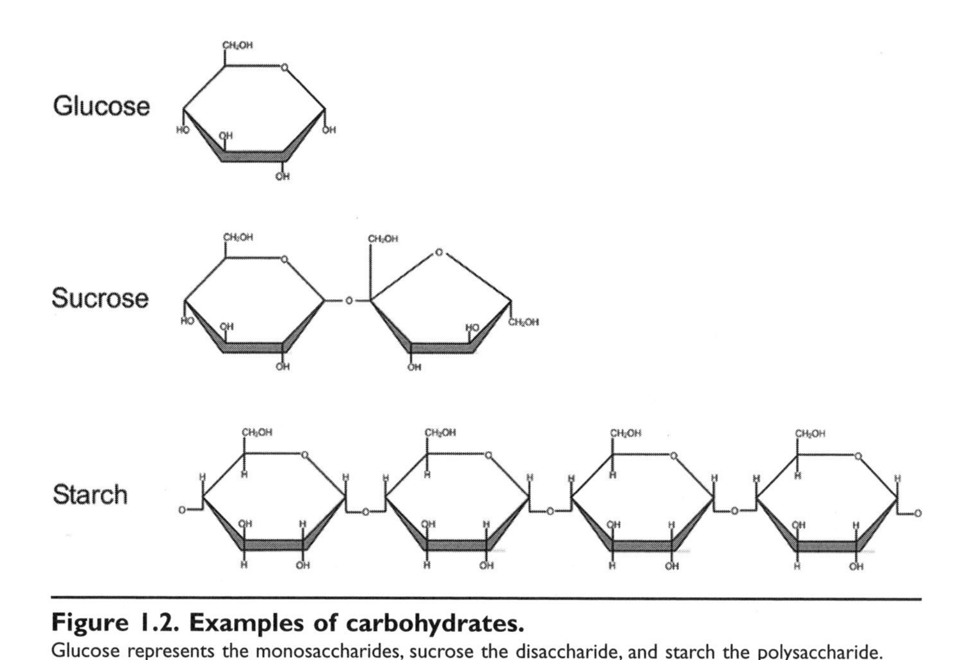

Figure 1.2. Examples of carbohydrates.
Glucose represents the monosaccharides, sucrose the disaccharide, and starch the polysaccharide.

detail in Chapter 10. Together, the monosaccharides and disaccharides all commonly called the *simple sugars*.

Complex carbohydrates, or the *polysaccharides*, are composed of long chains of glucose units. The different classes of polysaccharides vary in the physical structure of the chemical bonds that link the glucose units. In some cases, such as starch, the chemical bonds are easily digested by the human digestive system and thus provide a useful source of glucose for energy. However, a slight change in the configuration of the chemical bonds between the glucose units makes the bonds inaccessible by human digestive enzymes.

These molecules are called *fibers*, and even though they are not digestible by human enzymes, they play an increasingly important role in human digestion. Since they are basically indigestible, fibers provide bulk to food. This bulk helps move materials through the system and provides resistance to muscles of the gastrointestinal tract. This resistance acts as a form of workout for the muscles, which keeps them strong, allowing for the efficient movement of food in the system. Fibers exist in one of two general categories: soluble and insoluble. The soluble fibers, which are readily dissolved in water, are found primarily in fruits. These fibers slow down the movement of food through the gastrointestinal tract as well as the absorption of glucose. In contrast, the insoluble fibers, which are found in bran material and whole grains, increase the rate at which material is moved through the gastrointestinal tract, as well as provide bulk to the fecal material. The processing of fiber in the lower gastrointestinal tract will be covered in greater detail in Chapter 3. The additional role of fibers in preventing diseases of the gastrointestinal tract such as colon cancer and diverticulitis will be discussed in Chapter 10.

Fats and Lipids

While the carbohydrates are regarded as short-term energy sources for the human body, the fats and lipids are involved with more long-term energy processes within the body. There are exceptions to this, but in general the fats and lipids take more time to process by the digestive system and are associated with developing the long-term energy stores of the body. Technically, the term *lipid* is used to represent the entire class of these molecules, with the term *fat* primarily being reserved for a group of lipids called the *triglycerides*. However, frequently in nutritional analysis and on consumer products the terms are used interchangeably. In this volume the term *fat* will be reserved for the triglycerides, with *lipids* indicating the entire class of molecules. There are two major classes of lipids that are of interest in understanding the physiology of the digestive system of humans: the triglycerides and the sterols. A third class, the **phospholipids**, plays an important role in the structure of cell membranes.

For the most part the lipids are **hydrophobic** molecules, meaning that they do not dissolve readily in water. Because the digestive system is a water environment, as is the circulatory system, this physical characteristic of the lipids means that the digestive system will have to handle the lipids differently than most other nutrients. Chemical secretions such as bile, and specialized proteins called the **lipoproteins**, will assist the processing and transport of these important energy nutrients. This will be covered in detail in Chapter 2.

The triglycerides (see Figure 1.3) make up the majority (95 percent) of the lipids in food. The structure of these molecules, with its high number of carbon-carbon bonds, makes them an excellent source of energy for aerobic respiration. The long chains, called fatty acids, can vary in length and their degree of **saturation**. It is these chains that make the triglycerides hydrophobic. The level of saturation of the fatty acid chains has also been associated with human health. Saturated fats, found typically in animal products, are known to increase the risk of heart disease, while the unsaturated fats of plant products produce less of a risk. While the words *triglyceride* and *fat* have a negative connotation in today's society, in fact they are necessary and useful molecules in human metabolism when they are consumed in correct quantities. Some fats, called the omega-3 and omega-6 fatty acids due to their structure, actually help regulate the lipid biochemistry of the blood. Fats and lipids also provide insulation for the body. Because the fats give texture to food, release a pleasing aroma when cooked, and provide a fullness to the meal, they can be easily over-consumed during eating.

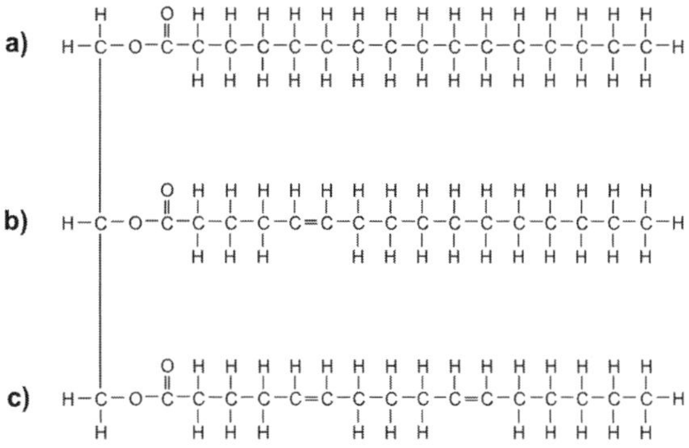

A second major class of lipids is the *sterols*, of which the most common is called cholesterol. As is the case with the triglycerides, the word **cholesterol** does not have a positive image in today's society. However, just like the triglycerides, cholesterol is an important molecule for our bodies. It serves as the starting material for the manufacture of important **hormones** such as testosterone and estrogen, is a component of the membranes of our cells, is used by the liver to manufacture bile (see Chapter 4), and is the starting material for the synthesis of vitamin D in our bodies. In fact, cholesterol is such an important molecule to our bodies that our liver has the capability of manufacturing all of the body's daily requirements of cholesterol.

Figure 1.3. A triglyceride.
The fatty acid chain at the top (a) represents a saturated chain. The middle chain (b) is monounsaturated and the lower chain (c) is polyunsaturated.

Like the triglycerides, cholesterol is a hydrophobic molecule, and thus the body is going to have some problems moving it around. To remedy this, cholesterol (and triglycerides) are packaged into a group of special transport molecules called lipoproteins. (This process will be discussed in greater detail in Chapter 2.) You should think of these lipoproteins as balloons. When the balloons are empty of cholesterol, they are compact and small and are called high-density lipoproteins (HDL). When there is an abundance of cholesterol in the system, the balloon is full and they are called low-density lipoproteins (LDL). Unfortunately, these lipoproteins have been given the names "good" (HDL) and "bad" (LDL) cholesterol, but that really is not correct. Because the body has the ability to manufacture all of its cholesterol, an overabundance in foods, called dietary cholesterol, will fill the balloons creating LDLs. A diet low in fat and cholesterol will leave the balloons empty, which are the HDLs. Scientists and nutritionists are still debating the effects of dietary cholesterol on the body. The amount of cholesterol does not directly influence the operation of the digestive system (although the foods that cholesterol is associated with do), but it does affect the health of the circulatory system.

There are additional classes of lipids, such as the waxes and phospholipids. Although important to the overall operation of the human body, they have little influence on the physiology of the digestive system and are processed in the same manner as the triglycerides and sterols.

Proteins

While the body primarily uses lipids and carbohydrates for energy, proteins are involved in a wide variety of functions other than supplying energy. To put it simply, proteins are the working molecules of the cell and, as such, are involved in the majority of all cellular functions. Proteins may have structural functions, such as those found in muscles; others work as signaling molecules in the nervous system or as hormones for the endocrine system. This list of protein functions in the human body is almost too numerous to mention, but a specialized group of proteins, called the enzymes, are an important component of the digestive system and are covered in the next section.

The building blocks of proteins are the **amino acids**. There are twenty different amino acids that are needed to construct the proteins of the human body. Each of these amino acids has a slightly different molecular structure, which gives each of them a unique chemical characteristic (see Figure 1.4). Within each cell, using instructions from the genetic material (deoxyribonucleic acid,

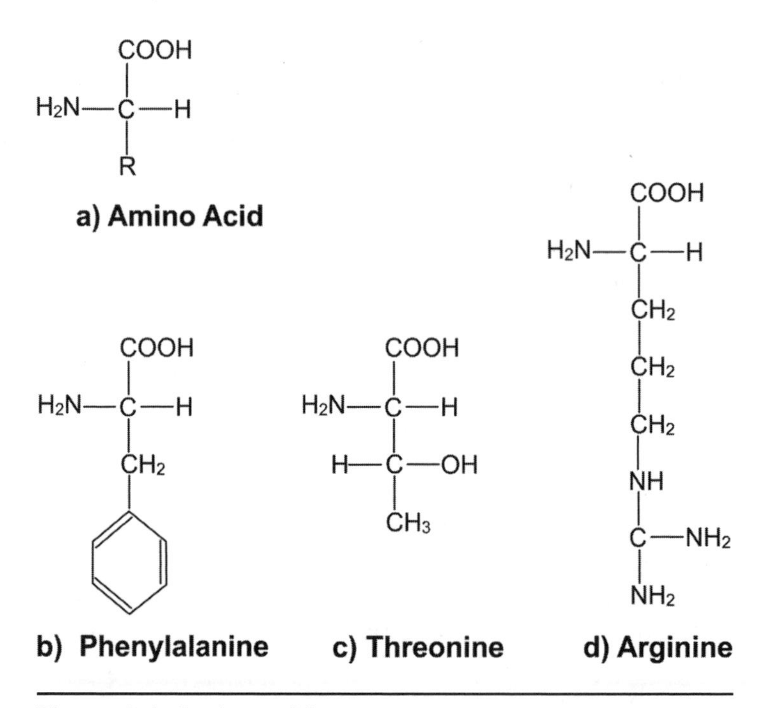

Figure 1.4. Amino acids.
This figure gives an example of several amino acids necessary for the formation of proteins. The top structure (a) represents the generic structure of an amino acid. The R represents a variable functional group. Phenylalanine (b) is an example of an aromatic amino acid, threonine (c) is an example of a polar amino acid, and arginine (d) is an example of a basic amino acid.

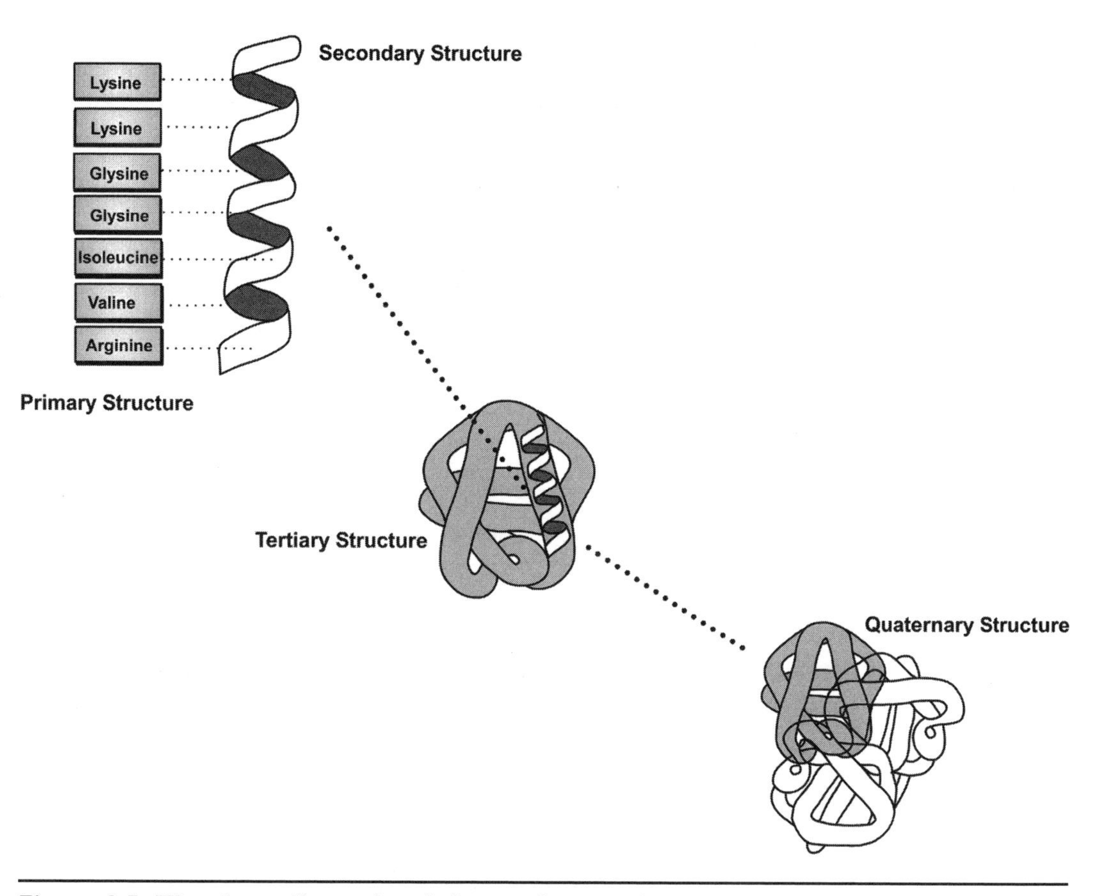

Figure 1.5. The three-dimensional shape of a protein.
This diagram indicates the four levels of protein structure for a basic globular protein. Primary structure is the linear arrangement of amino acids. Secondary structure represents the first level of protein folding. Tertiary structure develops the final three-dimensional shape of the protein. Some proteins form complex arrangements with other proteins. This is called quaternary structure.

or DNA), amino acids are linked together by the process of dehydration synthesis to form proteins. Like other molecules, these bonds are broken by the process of hydrolysis. However, unlike the carbohydrates, when the amino acid chains are formed, they fold into complex three-dimensional shapes (see Figure 1.5) that inhibit digestion in the body. Furthermore, the peptide bonds that hold the amino acids together are exceptionally strong, thus requiring the assistance of enzymes to break them down.

Because the proteins in food are the product of the original organism's body, and were constructed by the organism for a specific purpose, few proteins that are brought into the human digestive system are usable in their current form. Instead of absorbing whole proteins into the body, the role of the digestive system is to break down the protein bonds into their amino acid building blocks. The individual amino acids are then absorbed into the body and used as raw materials for the building of human-specific proteins. The digestion and absorption of proteins will be covered in more detail in Chapter 3.

Excess protein in the body can be used in the aerobic respiration pathways of the cell for energy and for this reason they are technically considered to be energy nutrients. However, before using the excess proteins the liver must remove the nitrogen from the amino acids. This process, called **deamination**, produces large amounts of urea and for this reason an elevation in dietary protein may place a strain on both the liver and urinary systems. The value of high-protein diets is still debated amongst nutritionists and doctors alike.

Enzymes

For the most part, the metabolic reactions of the body, including the dehydration synthesis and hydrolytic reactions mentioned previously, do not occur spontaneously. Instead, they require a catalyst to accelerate the rate of the reaction to a point that is efficient for the cells of the body. These catalysts are called enzymes. While the human body has a wide array of enzymes, which control everything from the operation of the nervous system to the process of cell division, they all share some common characteristics. First, the vast majority of all enzymes are proteins. The three-dimensional shape of enzymatic proteins enable them to interact with other molecules. Second, enzymes are very specific to the molecules, or substrates, that they interact with. Third, enzymes all serve to increase the efficiency of metabolic reactions by lowering the amount of energy needed to initiate the reaction. Finally, enzymes themselves are not consumed or destroyed during the course of an enzymatic reaction, allowing them to be reused over and over again for the same process.

The activity of an enzyme may be regulated by a variety of mechanisms. First, enzymes all have a specific environment in which they are the most efficient. The temperature and pH (or acid/base level) of the enzyme's environment act as a switch to regulate the activity of an enzyme. Since within the digestive system of humans the temperature remains a relatively constant 98.6°F (37°C), digestive enzymes are primarily regulated by the **pH** of their environment. The level of compartmentalization in the human digestive system helps to establish zones of enzyme activity. Throughout the next several chapters we will examine how the stomach and small intestine regulate the pH of their environments to control enzyme activity.

The digestive system utilizes a large number of enzymes to break down the nutrients within food into units small enough to be transported by the circulatory or lymphatic system. In this volume, we will usually refer to the enzymes by their general function. For example, enzymes that aide in the processing of lipids are called lipases, and those that process proteins are called proteases. See Table 1.1 for a list of the general enzyme class for each energy nutrient. Note that the general enzyme of carbohydrates is called an amylase. The prefix *amyl-* means sugar. Other enzymes and their mechanism of regulation will be discussed in the next several chapters.

Vitamins, Minerals, and Water

The action of the digestive system is not confined solely to the processing of the energy nutrients. The human body requires a daily input of other nutrients to meet its metabolic requirements. The processing of vitamins, minerals, and water differs from that of the energy nutrients in that these nutrients are usually not broken down by the digestive system, but rather are absorbed intact and then transported by the circulatory system to the other systems of the body. While is it beyond the scope of this book to describe all of the vitamins and minerals, some basic characteristics of these nutrients are described in the following paragraphs so as to provide an overview of how these nutrients interact with the digestive system. The Organizations and Web Sites chapter and the bibliography of this book provide a list of useful sources for additional information on specific nutrients.

VITAMINS

Vitamins are similar to the energy nutrients in that they are organic molecules, but differ in the fact that the body does not get energy directly from these molecules. Instead, vitamins serve as enzyme assistants, or coenzymes. Some vitamins, specifically the B-complex vitamins, are directly involved in the processing of energy nutrients, specifically lipids and carbohydrates. Certain vitamins serve as protectors of the delicate cellular machinery. These are called the **antioxidants** and are best represented by vitamins C and E. Others aid in the vision pathways (vitamin A), or in the building of healthy bones (vitamins D and A). Nutritionists divide the vitamins into two groups based upon how they interact with the body. The first are the water-soluble vitamins, a group that consists of vitamin C and the B vitamins (see Table 1.2). These vitamins are readily absorbed by the digestive system and, with a few exceptions, do not require special processing. The other class, known as the fat-soluble vitamins (vitamins A, D, E, and K), are frequently treated in the same manner as the triglycerides, meaning that they are packaged into specialized lipoproteins and transported by the lymphatic system. In general, both classes are required in relatively small quantities (micrograms or less) daily by the body.

There are two vitamins that we will focus on in some detail in this volume. The first is vitamin D, which is produced by the body using the cholesterol in the skin as a starting material. When the skin is exposed to

TABLE 1.2. Vitamins and Minerals

Water Soluble Vitamins	Fat Soluble Vitamins	Major Minerals	Trace Minerals
Vitamin C	Vitamin A	Sodium	Iron
Thiamin	Vitamin D	Potassium	Zinc
Riboflavin	Vitamin E	Calcium	Copper
Niacin	Vitamin K	Chloride	Manganese
Pantothenic acid		Phosphorous	Iodine
Vitamin B_6		Magnesium	Fluoride
Vitamin B_{12}		Sulfur	Chromium
Folate			Molybdenum

This table lists the different classes of vitamins and minerals needed by the human body. The trace mineral category contains a list of the most common minerals in this class. Scientists are investigating the role of other minerals, such as nickel, in human metabolism.

sunlight, specifically ultraviolet radiation, the chemical structure of the cholesterol is modified to create a precursor of vitamin D. This chemical is then transported to the liver and adrenal glands for additional processing. Vitamin D acts like a hormone, in that it regulates the calcium absorption properties of calcium in the small intestine. The role of vitamin D in this process is discussed in greater detail in Chapter 3.

The second vitamin of interest in the study of the digestive system's physiology is vitamin K. Vitamin K is a vitamin that is involved in a wide-variety of body functions, most notably the clotting response of the blood. Some vitamin K is produced by the naturally existing bacteria of the large intestine, or *colon*. The role of these bacteria in vitamin K formation and establishing a healthy environment in the colon is covered in Chapter 3.

As with many of the nutrients, there is a significant amount of misinformation in the popular media regarding the ability of some vitamins to prevent disease, enhance performance, increase memory, and so on. The Organizations and Web Sites chapter and bibliography of this volume provide a reliable list of reference materials for those who are interested in learning more accurate information about vitamins.

MINERALS

Minerals are inorganic nutrients that play an important role in the regulation of many of the body's metabolic functions. Like the vitamins, many minerals function as assistants to metabolic pathways. Still others help regulate body fluid levels, and some serve as structural components of bones.

Minerals are also the major **electrolytes** in the circulatory system. Nutritionists divide minerals into two broad classes: the trace minerals and the major minerals (see Table 1.2). It is important to note that the terms *trace* and *major* do not reflect the importance of the mineral in the body, but rather the abundance of the mineral in the human body. For example, iron is considered to be a trace mineral, but it is crucial to the development of hemoglobin in the blood.

The digestive system handles minerals in a variety of ways. Some minerals, such as sodium and potassium, are quickly absorbed from food and transported by the circulatory system. However, some minerals, such as calcium, are poorly absorbed by the gastrointestinal (GI) tract. This level of potential availability of minerals is frequently called *bioavailability* and reflects not only the physical interaction of the digestive system with the mineral, but also the presence of certain chemicals in foods that may bind the mineral and make it unavailable to the digestive system. In addition, the ability of the GI tract to extract minerals from food is dependent on the overall health of the system, the age of the person, their sex, and other factors, such as pregnancy. The role of some of the more important minerals, such as iron, calcium, sodium and potassium, will be discussed throughout this volume. By examining these minerals in detail, one can gain an appreciation of how the GI tract processes minerals in general. For additional reading on the roles of the minerals and their food sources, please consult Organizations and Web Sites chapter and the bibliography of this book.

WATER

While the average person may not consider water to be a nutrient, in fact it is probably one of the most important nutrients for the digestive system. Like the circulatory, respiratory and urinary systems, the digestive system is a water-based system that uses water to move nutrients, deliver digestive enzymes, lubricate the length of the gastrointestinal tract, and facilitate the absorption of nutrients into the circulatory and lymphatic systems. The average human requires about 2.65 quarts (approximately 2.5 liters) of water per day to meet the metabolic requirements of the body. The majority of this comes from liquids and foods that are consumed throughout the day. A smaller amount is derived from chemical reactions within the body, such as dehydration synthesis reactions.

The movement of water between the digestive system and the tissues of the body, most commonly the circulatory system, is highly regulated. The digestive system must simultaneously retain enough water for its own operation and supply the body with the water it needs to function. This is a complex task and frequently involves the use of minerals such as potassium and sodium to establish concentration gradients to efficiently move water. The large intestine, or colon, is the major digestive organ responsible for this process, and its role in water regulation will be discussed in detail in Chapter 3.

The next chapters will outline the anatomy and physiology of the upper and lower sections of the gastrointestinal tract. Although these chapters will examine the action of the individual organs of the digestive system, at the cellular level the digestive system is based on the cellular and molecular interaction of enzymes and other chemicals with the nutrients of food.

The Upper Gastrointestinal Tract: Oral Cavity, Esophagus, and Stomach

The human digestive system is actually a series of organs that form a long, enclosed tube. This organ system of the human body is specialized for breaking down incoming food into the needed nutrients for the body's vast array of metabolic functions. The majority of the organs in the human body are either directly or indirectly associated with the process of digestion. The organs of the GI tract are those that physically comprise the tube, also called the *alimentary canal*, which the food physically passes through. These include the oral cavity, esophagus, stomach, small intestine, and large intestine (also called the colon). Figure 2.1 gives the location of these organs in the human body. Associated with the organs of the gastrointestinal tract are the accessory organs. The accessory organs, which will be covered in detail in Chapter 4, contribute needed materials for the breakdown and processing of the food entering the system. In many cases the accessory organs have multiple functions that are highlighted in other volumes of this series.

For convenience, the GI tract is frequently divided into two major sections for study. The upper GI tract consists of the oral cavity, esophagus and stomach, as well as associated valves and accessory organs. The lower GI tract, which will be covered in more detail in the next chapter, consists primarily of the small intestine and colon. This division, while practical from the standpoint of a reference book, also has some basis in physiological function. As this chapter will explore, the role of the upper GI tract is primarily in the processing of food material. The majority of the digestion and

A young girl eating an apple. © Skjold Photographs.

nutrient processing, as well as the preparation of waste material, occurs in the lower GI tract, which will be covered in the next chapter.

THE ORAL CAVITY

The human mouth, also called the oral cavity or **buccal cavity**, is the entry point into the human digestive system. The oral cavity represents an area of intense activity for the body. Not only are nutrients initially processed in this location, but is also serves as the connecting point between the respiratory system and the outside environment (see Figure 2.2), as well as the location of a significant amount of sensory input from chemical receptors, most notably taste. The human mouth is the site of both mechanical and enzymatic digestive processes.

Salivary Glands

Vital to the digestive functions of the oral activity are the secretions of three pairs of accessory glands collectively called the salivary glands. These glands are identified by their location in the oral cavity. Two pairs are located along the bottom of the oral cavity. The *sublingular* glands are located just below the tongue and the *submandibular* glands are positioned just beneath these, near the *mandibula* (jaw bone). A third set, called the *parotid* glands, are located just in front of, and slightly below, the ears. The childhood disease called **mumps** frequently infects the parotid glands, although the other pairs may become infected as well. (See Chapter 11 for more information on mumps.) A duct carries the secretions of each salivary gland into the oral cavity.

Saliva, the chemical secretion of the salivary glands, is actually a complex mixture that performs a variety of functions for the digestive system. Saliva is primarily water (99.5 percent), which serves to lubricate and moisten the digestive system. However, it is the remaining 0.5 percent of the volume that contains some of saliva's most important functions. This small fraction contains important **ions**, such as potassium, chloride, sodium, and phosphates, which serve as pH buffers and activators of enzymatic activity.

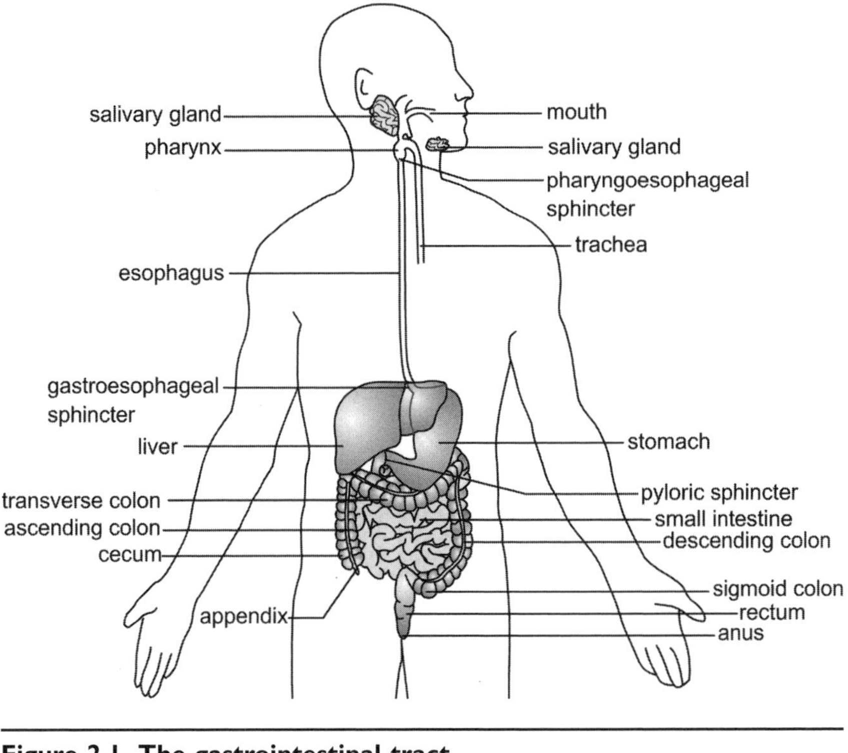

Figure 2.1. The gastrointestinal tract.
This diagram gives the general arrangement of the digestive and accessory organs.

Since salivary glands are similar in structure to sweat glands found within the skin, they also secrete urea and uric acid as waste products. Saliva also contains a small amount of an enzyme called *lysozyme*, which inhibits, but does not eliminate, the formation of bacterial colonies in the oral cavity. *Mucus*, a watery mixture of complex polysaccharides, helps lubricate and protect the oral cavity. Also found in saliva is another enzyme, called *salivary amylase*, which initiates the process of carbohydrate digestion (discussed at length in the following section).

The composition of the saliva varies slightly depending on the salivary gland in which it originates. The salivary glands of an adult can secrete a combined volume of 1.58 quarts (1,500 milliliters) of saliva daily. The amount of saliva secreted at a specific time is dependent on a number of factors. For example, when the body is dehydrated, the production of saliva is decreased, which in turn contributes to a thirst response by the body. An increase in saliva production is under direct control of the brain and is usually the result of a response to a chemical stimulus. The sight or smell of food typically serves to increase saliva production. Memories can also result in an increase of saliva production, such as the memory of a favorite

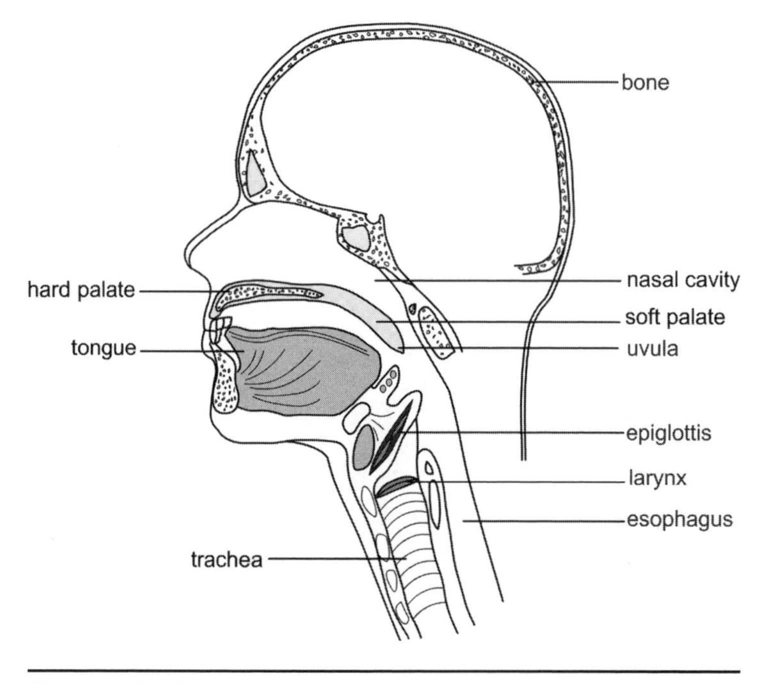

Figure 2.2. The structures of the oral cavity, including the relationship to the nasal cavity and respiratory passages.

food or food-related event. Increased saliva production can continue for some time after eating to cleanse the mouth of food, eliminate harmful bacteria and restore the normal pH of the oral cavity.

Mechanical Digestion

In the oral cavity the process of mechanical digestion serves several functions. First, the action of the teeth and tongue break the food into small portions so that it may be sent to the stomach via the esophagus. Second, the process of mechanical digestion increases the surface area of the food, allowing the secretions of the salivary glands to mix freely with the food and stimulating the action of the taste buds.

The action of chewing, or mastication, is the first stage of mechanical digestion. Chewing involves the action of both the teeth and the tongue. While there are three major types of teeth in a human adult mouth (molars and premolars are frequently classified as one type; see Figure 2.3), all teeth have the same fundamental structure (see Figure 2.4). It is the shape of the tooth that determines its function in mechanical digestion (see Table 2.1). The combination of different types of teeth in the mouth allows for the processing of a large variety of foods, from protein-rich meats to nutritious veg-

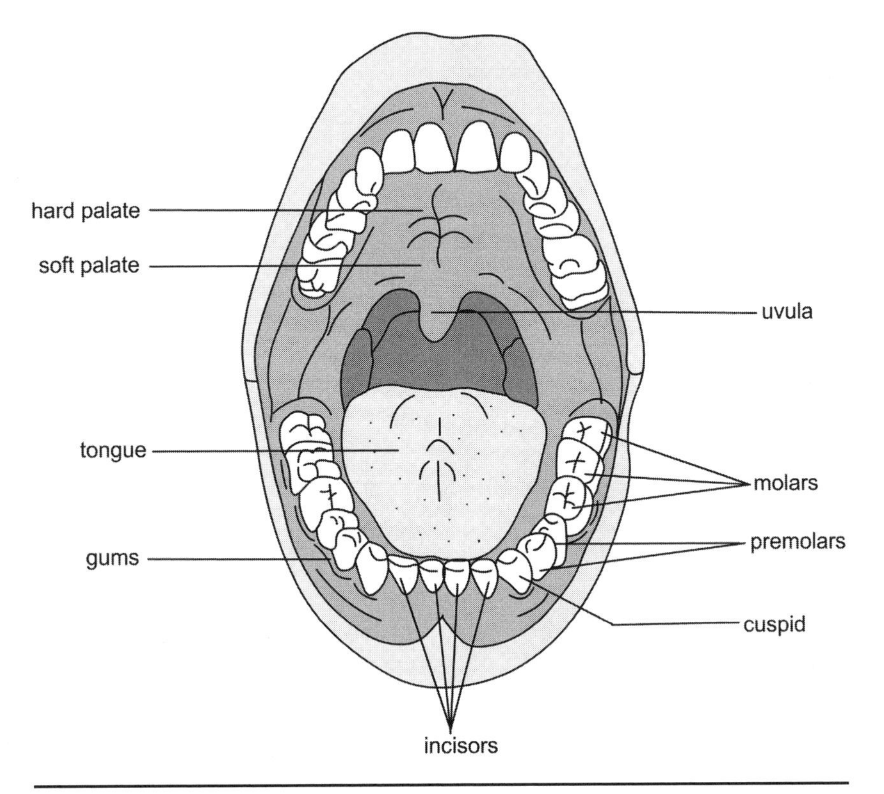

Figure 2.3. The oral cavity showing the location of the major structures and the position of the types of human teeth.

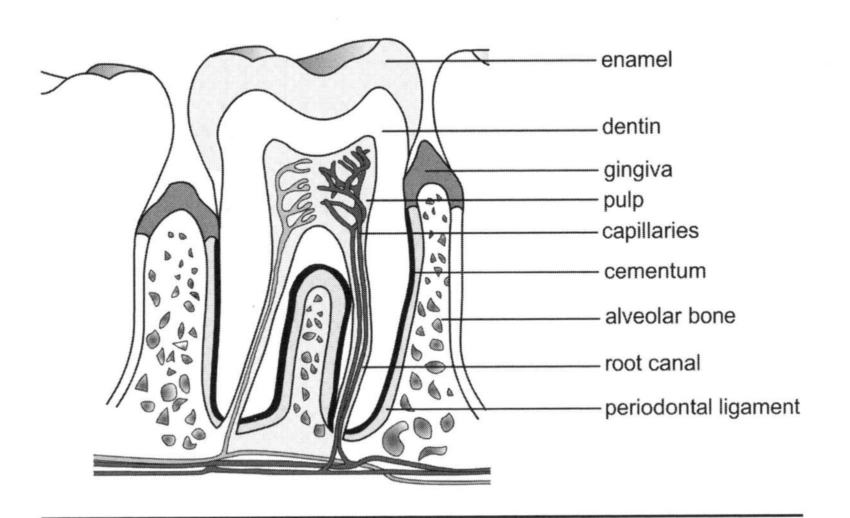

Figure 2.4. The internal structure of a typical human tooth.

TABLE 2.1. The Teeth of an Adult Human

Class	Function	Number
Incisors	cutting food	8
Cuspids (canines)	tearing and shredding	4
Molars	crush and grind food	
Premolars		8
Molars		12

etables and fruits. The shape and structure of the human jaw is designed to provide a large physical force to the teeth, which can be used to grind both plant tissue and bones to release nutrients.

The teeth absorb the brunt of this force. To prevent damage, each tooth is located in a socket of the jaw bone. Connecting each tooth to the socket is the *periodontal ligament*, which also acts as a shock absorber. The socket and lower portions of the tooth are covered by the gums, or *gingivae*. (A common inflammation of this tissue is called *gingivitis*.) Teeth are made from a calcified form of connective tissue called **dentin**, which is covered with a combination of calcium phosphate and calcium carbonate commonly called *enamel*. (Dental caries typically erode this area of the tooth.) Within the center of each tooth is an area called the *pulp cavity*, which contains nerves, blood vessels and ducts of the lymphatic system. (The diseases of the teeth, including cavities and periodontal disease, will be discussed in Chapter 8.)

Without teeth, humans would be required to swallow food whole in much the same manner as snakes. As organisms with a high metabolic rate, we require a relatively rapid processing of incoming nutrients. The importance of the teeth in increasing the surface area of the food for later enzymatic digestion should not be underestimated.

The second stage in the mechanical processing of the food involves the action of the tongue. The tongue is comprised of **skeletal muscle**, which is under the voluntary (but not always conscious) control of the body. The movement of the tongue is controlled by two separate sets of muscles. The *extrinsic muscles* enable the movement of the tongue that is important for digesting food. These muscles move the food from the area of the teeth to the back of the mouth, where it is formed into a small round mass of material called a **bolus**. This area at the rear of the oral cavity is commonly called the **pharynx**. The pharynx serves as the junction between the respiratory system and digestive system, and thus all activity in this area must be highly coordinated by the body. By the action of the tongue, the food is lubricated with saliva to facilitate swallowing and to mix in the enzymes of the salivary glands. The tongue also participates in the swallow reflex (see the following section) through the action of the *intrinsic muscles*. This muscle group also controls the size and shape of the tongue and is involved with speech.

Located on the tongue are a series of **papillae**, which are small projections of the tissue. It is the papillae that give the tongue its rough texture. The papillae are sometimes mistakenly referred to as the *taste buds*, but the taste buds are actually specialized receptors located at the base of certain types of papillae. There are three different forms of papillae, which differ in their appearance and location on the tongue. The *circumvallate papillae* are the largest and are located in a V-shaped region at the rear of the tongue (see Figure 2.5). All of the circumvallate papillae contain taste buds. The *fungiform papillae* are knoblike in appearance and are dispersed across the entire tongue. Depending on their location, some fungiform papillae contain taste buds. The last group is the *filiform papillae*. These have a filament-like appearance and are also uniformly distributed across the surface of the tongue. However, unlike fungiform papillae, filiform papillae rarely contain taste buds.

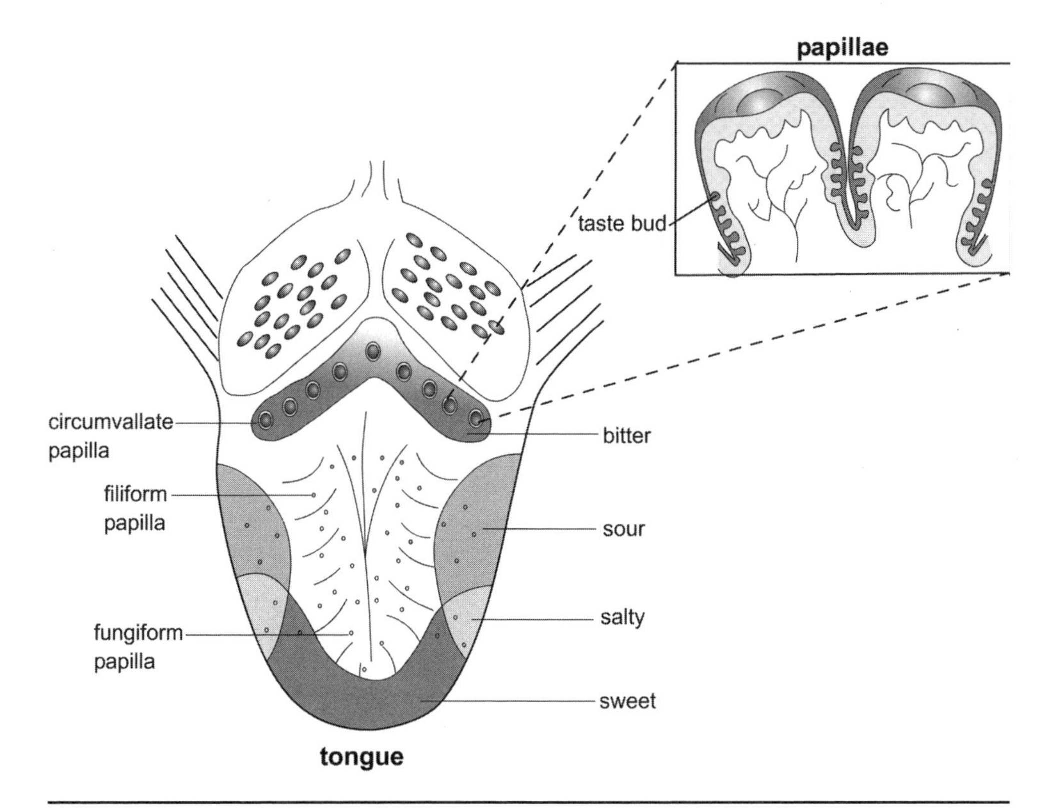

Figure 2.5. The external structure of the tongue showing the relationship between papillae types and taste zones.

The tongue is divided into four different taste zones—sour, bitter, sweet and salty (Figure 2.5). The taste buds in each area are sensitive to a unique chemical signature. Depending on the origin of the signal, the brain interprets the different tastes. The number of receptors that fire and the duration of the signal determines the intensity of the taste.

Enzymatic Digestion

As noted in the first chapter, enzymatic digestion is responsible for breaking organic material into smaller subunits that can be absorbed into the circulatory system. The amount of enzymatic digestion within the oral cavity is small in comparison to the activity of the lower GI tract. However, there is some initial digestion of both carbohydrates and lipids in the oral cavity.

The salivary glands, primarily the submandibular and sublingual glands, secrete an enzyme called salivary amylase. Recall from the first chapter that the nutrients are primarily absorbed from the digestive system in their simplest structure, or monomers. Salivary amylase belongs to a class of enzymes that digest complex carbohydrates, such as starch, into monosaccharides. The monosaccharides are easily absorbed into the circulatory system, although little absorption occurs in the oral cavity. The salivary amylase is mixed into the food by the action of the tongue and cheeks and continues to break down the starches in the food for about an hour until deactivated by the acidic pH of the stomach. A second enzyme of the oral cavity is *lingual lipase*. Lingual lipase is secreted from glands on the surface of the tongue. This enzyme acts on triglycerides in the food, breaking them down into monoglycerides and fatty acids. However, the action of this enzyme is relatively minor and it does not make a major contribution to overall lipid digestion.

Swallowing Reflex

As the bolus forms in the rear of the oral cavity, or pharynx, the swallowing reflex begins. Swallowing, or **deglutition**, is a staged process that is in part under voluntary control and partially a reflex action. While most people do not consciously think of swallowing, in fact it represents a complex, highly coordinated activity. The tongue, through the action of the intrinsic muscles, forces the food to the back of the mouth. The pressure of the bolus on the pharynx activates a series of receptors that send a signal to the swallowing center of the **medulla oblongata** and **pons** in the brain. The swallowing center then temporarily deactivates the respiratory centers of the brain to ensure that the bolus will be directed into the digestive, and not the respiratory, system.

As the bolus prepare to enter the esophagus, a series of events is initiated to direct the food into the digestive system. Once the swallowing reflex has begun the following four events occur in rapid succession (see Figure 2.6):

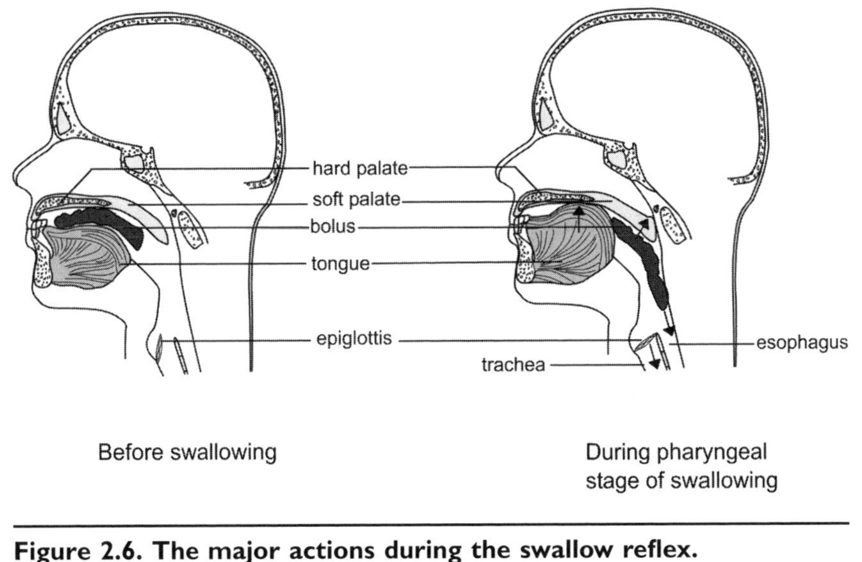

Figure 2.6. The major actions during the swallow reflex.
The dark arrows indicate the movement of structures during the process of swallowing. Notice that during swallowing the nasal cavity and respiratory pathways are closed.

1. The tongue moves upward against the roof (hard palate) of the mouth to prevent the food from reentering the oral cavity.
2. The *uvula*, an inverted-Y shaped flap of skin at the rear of the mouth, moves upward to block the nasal passages.
3. The vocal cords in the larynx tightly close over the opening of the windpipe, or *glottis*.
4. As the bolus passes into the esophagus, it forces a flap of cartilaginous tissue called the *epiglottis* downward over the glottis as an added precaution to protect against the food entering the respiratory system.

LAYERS OF THE DIGESTIVE SYSTEM

Before following the bolus on its brief journey through the esophagus, it is necessary to discuss the tissue structure of the digestive tract. From the esophagus to the anus, the walls of the digestive tract have the same general structure, with minor variations in each organ to enable specific functions. Within the wall of the digestive tract are four major tissue layers. From outermost to innermost they are the *serosa*, *muscularis externa*, *submucosa*, and *muscosa* (see Figure 2.7).

The serosa is the outermost layer of the digestive tract and is comprised of connective tissue. The serosa is important in that it forms a connection between the digestive tract and the **mesentery** that suspends the organs of the digestive tract within the abdominal cavity. To prevent friction between

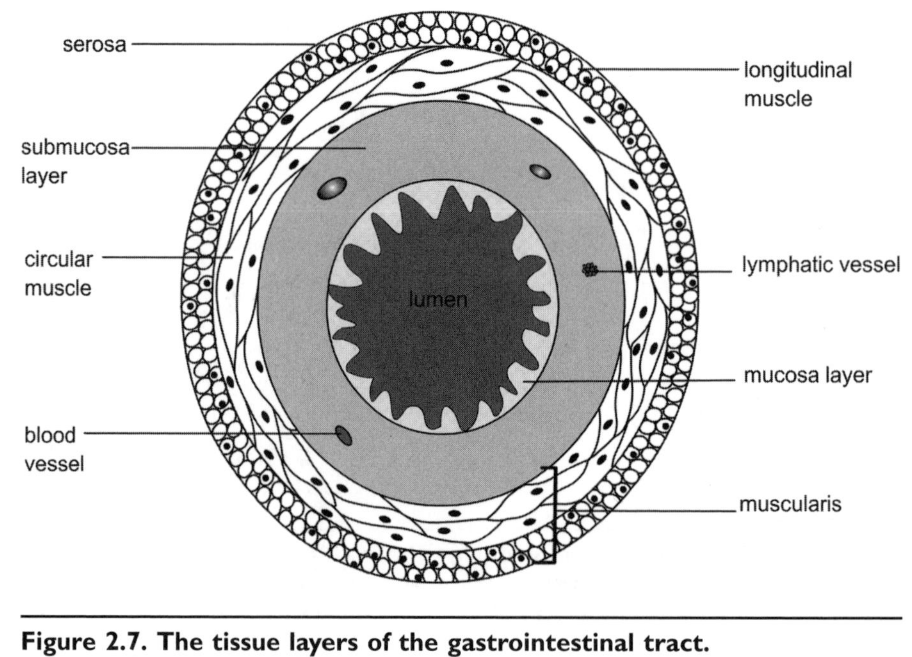

Figure 2.7. The tissue layers of the gastrointestinal tract.
The lumen represents the internal cavity of the gastrointestinal tract.

the organs of the system, the serosa secretes a water-based mixture that lubricates the exteriors of the organs. Directly under the serosa is a double layer of smooth muscle, the muscularis externa. These two muscle layers are the inner circular muscle and the outer longitudinal muscle. Since these layers are composed of smooth muscle, they are not under voluntary control of the brain. However, a nerve network called the *myenteric plexus* allows for regulation of activity from the involuntary control centers of the brain. The muscles contract in different directions, with the circular layer controlling the diameter of the digestive tract and the longitudinal layer controlling the length. The human digestive system does not rely on gravity to move nutrients through it, instead it is the action of these two muscle layers that rhythmically moves food through the system by a series of coordinated contractions called **peristaltic action**.

The next layer inward is a dense section of connective tissue called the submucosa. Located within the submucosa are the major blood and lymphatic vessels, as well as another series of nerves called the *submucous plexus* that provides involuntary regulation of the layer. The innermost layer is the mucosa. It is this layer that lines the interior of the digestive tract and thus is in direct contact with the nutrients passing through the system. The **epithelial cells** of the mucosa serve several functions, depending on the region of the gastrointestinal tract. In some cases these cells secrete a mucus layer that serves to lubricate the passage and protect the cells. Other cells may secrete digestive juices, while still others may release hormones that regulate the activity of the region. Epithelial cells are arranged into folds to increase the surface area. The amount of folding is dependent on the region of the gastrointestinal tract (for an example see the section of Chapter 3 on the small intestine). Also within the mucosa, usually just underneath the epithelial cells, is a thin layer of smooth muscle, as well as blood vessels, lymphatic vessels, nerves, and cells of the immune system.

ESOPHAGUS

Once the bolus leaves the oral cavity it enters into a muscular tube called the esophagus (see Figures 2.2 and 2.6). The esophagus is not a major digestive organ because the only enzymes that are active here are the salivary amylase and lingual lipase from the oral cavity. Furthermore, since the bolus only spends a brief amount of time in the esophagus (between five and nine seconds) and the mucosa tissue layer does not contain a large number of folds, there is almost no absorption of nutrients through the walls of the esophagus. Instead, the esophagus serves as a conduit from the oral cavity, through the thoracic region of the body, and to the stomach. The thoracic region houses the heart and lungs and is bordered on the bottom by a muscular barrier called the **diaphragm**. The esophagus passes through the diaphragm and connects to the upper portion of the stomach.

The bolus moves through the esophagus by peristaltic action. To aid the movement of the bolus, the cells of the mucosa tissue layer secrete mucus to lubricate the tube. To ensure the one-way movement of food, the esophagus is regulated by two sphincters, or valves. At the upper end of the esophagus is the *pharyngoesophageal sphincter*, which also serves to limit the flow of air into the gastrointestinal tract during breathing. At the lower end of the esophagus is the *gastroesophageal sphincter*, sometimes called the cardiac sphincter, which connects the esophagus to the stomach. The gastroesophageal sphincter also inhibits the reflux, or backup, of gastric juices from the stomach into the esophagus. Without this valve the highly acidic gastric juices would damage the delicate mucosa layer of the esophagus. (Medical ailments of the esophagus, including heartburn, are covered in more detail in Chapter 8; cancer of the esophagus is examined in Chapter 12.)

STOMACH

The stomach is commonly recognized as a muscular sac that functions as a holding site for food before it enters into the small intestine, as well as the location where the food is mixed and partially digested by me-

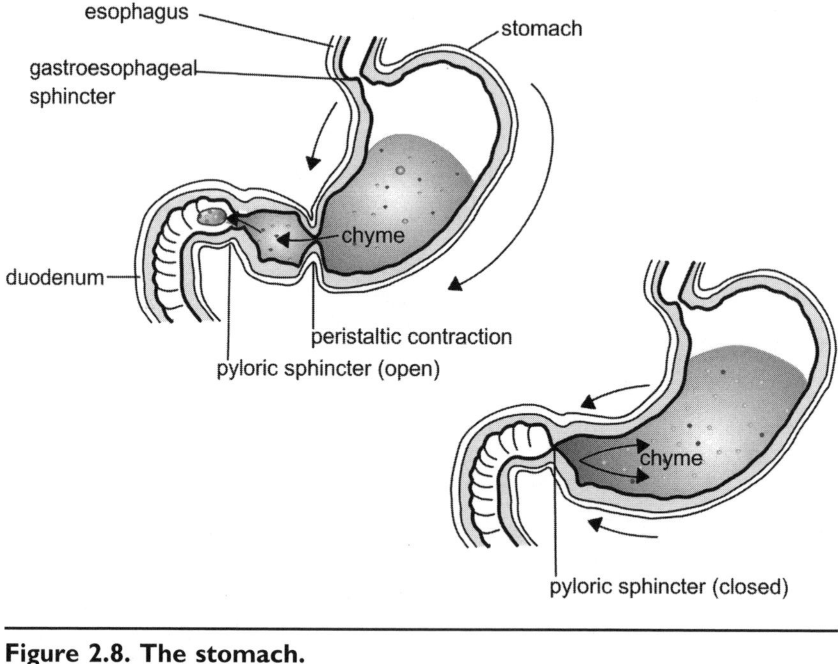

Figure 2.8. The stomach.
This diagram not only shows the location of the sphincters that define the boundaries of the stomach, but also the peristaltic contractions that are responsible for moving forward toward the duodenum.

chanical processes. However, while the stomach does perform these functions, in actuality its physiology and role in the digestive process is much more complex.

The stomach is an elastic, J-shaped organ whose boundaries are defined at the upper end by the gastroesophageal sphincter, and at the lower end by the *pyloric sphincter* (see Figure 2.8). When empty a human stomach may have a volume as little as 0.05 quarts (50 millimeters). In comparison, when full the stomach may contain almost 1.06–2.11 quarts (1–2 liters) of food, depending on the individual. The stomach contains the same tissue layers as are found in the esophagus, small intestine and colon, with some important variations in the secretions and structure of the mucosa layer. The J-shaped interior of the stomach is divided into regions based on slight differences in the secretions of the mucosa layer, thickness of the muscle layers, and overall function in digestion. The uppermost part, located above the level of the gastroesophageal sphincter, is the *fundus*. Below this is the main region of the stomach, called the *body*. The lower portion of the stomach, which connects to the small intestine, is called the *antrum*. A forth region, called the *cardia*, is located around the area of the gastroesophageal opening and plays only a limited role in digestion. The physiological differences of the fundus, body, and antrum will be covered within the following sections.

Composition of Gastric Juice

The stomach produces about 2.12 quarts (2 liters) of gastric juice per day. Recall that in the general structure of the digestive tract the mucosa tissue layer may contain folds. In the stomach these folds are called *gastric pits* (see Figure 2.9). Located in each of these pits are specialized cells that are responsible for generating the secretions of the stomach. At the top of the gastric pit are the *mucous neck cells*. Together with the surface epithelial cells, they are responsible for secreting the mucus coating that protects not only the cells within the gastric pit, but also the mucosa layer of the stomach. The mucus has an **alkaline**, or basic, pH, and thus serves to neutralize any stomach acid before it comes in contact with the stomach mucosa. The mucus also serves to lubricate the interior of the stomach for mechanical digestion (discussed later in this chapter). Located within the gastric pits are the *chief cells* and *parietal cells*. The chief cells secrete an inactive en-

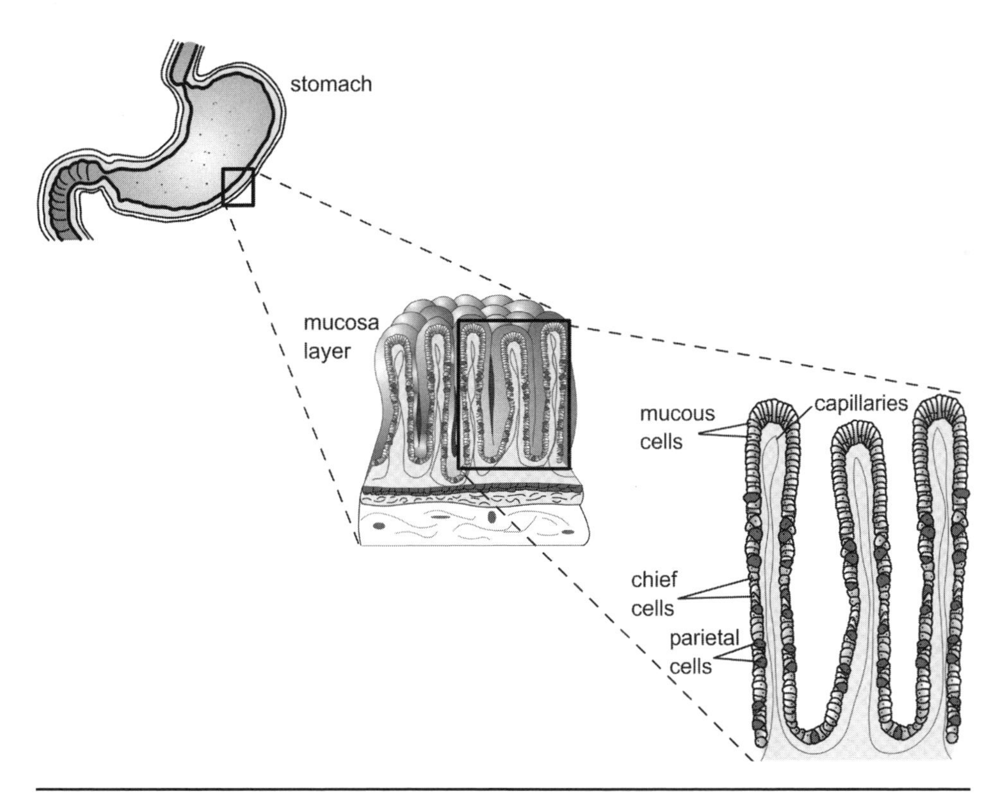

Figure 2.9. The gastric pits of the stomach showing the location of chief and parietal cells.

zyme called pepsinogen, which is involved in the chemical digestion of proteins (see next section).

The parietal cells are responsible for manufacturing hydrochloric acid. The manufacture of hydrochloric acid is an energy-intense process, and thus the parietal cells have an exceptionally high concentration of mitochondria to generate the needed energy. Hydrochloric acid has a pH of 2.0, making it 100,000 times more acidic than water. The hydrochloric acid serves a number of functions in the stomach. First, it distorts, or *denatures*, the structure of proteins, making them easier to digest. Second, the low pH of the hydrochloric acid activates the pepsinogen enzyme secreted by the chief cells. Finally, the low pH of the hydrochloric acid acts as a deterrent against bacterial contaminants in the food. In this regard the gastric juice of the stomach acts as a physical barrier of the immune system, although Chapter 9 will give examples of several organisms that have evolved mechanisms of avoiding or neutralizing its effects. In addition to hydrochloric acid, the parietal cells secrete an **intrinsic factor** that aids in the absorption of vitamin B_{12} in the small intestine.

While the gastric pits of the fundus, body, and antrum may look fundamentally the same, there are some important variations in the secretions from these areas. Gastric juice, containing hydrochloric acid, pepsinogen, some mucous, and intrinsic factors, is secreted primarily by the cells of the fundus and body. In comparison, the cells of the antrum are responsible for secreting a large amount of mucus. Specialized cells in this area, called G cells, release a hormone called *gastrin*. Gastrin is released directly into the bloodstream and regulates the activity of the parietal and chief cells in the body and fundus of the stomach.

Enzymatic Digestion

The food was mixed in the oral cavity with saliva, which contains salivary amylase. The chemical digestion of complex carbohydrates in the bolus continues down the esophagus. After the bolus passes through the gastroesophageal sphincter, and enters the stomach, it comes in contact with the highly acidic hydrochloric acid. While hydrochloric acid deactivates salivary amylase, the lack of a significant amount of mechanical digestion in the fundus and upper regions of the body of the stomach allows the salivary amylase to continue carbohydrate digestion within the bolus. However, once mechanical digestion begins lower in the stomach, the salivary amylase is quickly inactivated. The thick mucus coating of the stomach prohibits the absorption of digested carbohydrates into the bloodstream.

In the oral cavity, the enzyme lingual lipase initiated a limited digestion of triglycerides in the food. In the stomach, the chief cells also release a gastric lipase, which serves much the same function in breaking down triglycerides. As was the case with the carbohydrates in the oral cavity and stomach, this is not a major contribution to the overall digestion of these nutrients, and there is no appreciable absorption of triglycerides through the stomach lining.

The prime nutrient target of enzymatic digestion in the stomach is protein. Recall from Chapter 1 that proteins may be large molecules, and all contain multiple levels of complex organization. This three-dimensional structure of proteins, and the presence of peptide bonds holding the amino acids together, makes proteins a difficult class of nutrients to digest. The purpose of protein digestion in the stomach is to initialize the process by destabilizing the structure of the protein. Thus, the mechanisms of protein digestion in the stomach are very general, and not directed at one specific type of protein. As was the case with carbohydrates, there is no absorption of peptides or amino acids through the lining of the stomach.

Enzymes that are involved in protein digestion belong to the general class called *proteases*. The pepsinogen secreted by the chief cells in the gastric pits is initially inactive, so as to protect the cells of the gastric pit from unintentional digestion. After being secreted, the pepsinogen makes its way through the protective mucus coat and into the main cavity, or **lumen**, of the stomach (see Figure 2.10). The hydrochloric acid in the lumen activates the pepsinogen by cleaving off a small fragment from one end of the molecule. This active form of the enzyme is called *pepsin*. Pepsin also has the ability to activate pepsinogen in what is frequently called an **autocatalytic process**. Once activated, pepsin breaks down some proteins into smaller peptide fragments for further digestion later in the small intestine.

Mechanical Digestion

The stomach is primarily an organ of mechanical digestion, whose purpose is to thoroughly mix the incoming food material with gastric juice, forming a semi-solid mixture called *chyme*. This process occurs in three distinct stages: (1) the filling of the stomach with food and the temporary storage of food, (2) the mixing of the food with gastric juice, and (3) the emptying of the stomach. The purpose of these processes is to manipulate the chyme to the correct consistency, so that it can pass through the pyloric sphincter into the upper region of the small intestine, also called the *duodenum*, for digestion. A series of complex signals between the stomach and small intestine control the final movement of materials into the duodenum. As was the case with enzymatic digestion, each of the three regions of the stomach has slightly different roles in mechanical digestion.

After passing through the gastroesophageal sphincter, the bolus enters into the body of the stomach. With a normal volume of about 1.5 fluid ounces (50 milliliters), this region of the stomach would quickly fill with food if it were not for the elastic nature of the stomach lining. Located along the inside of the stomach are deep folds of tissue called *rugae*. The purpose

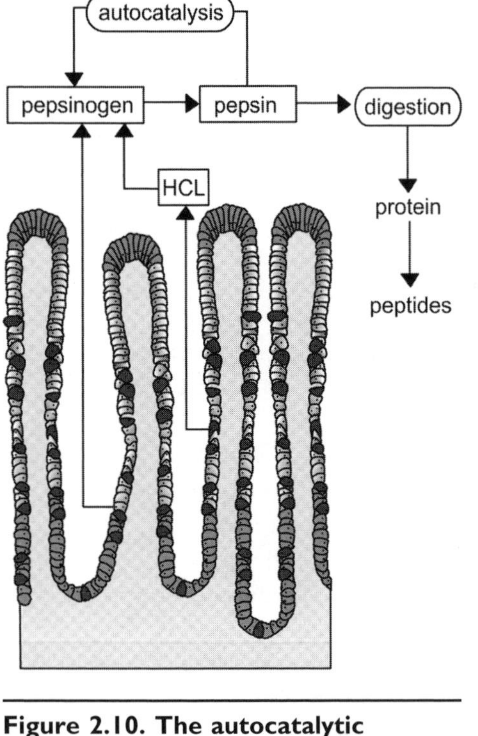

Figure 2.10. The autocatalytic activation of digestive enzymes in the stomach.

of the rugae is to allow the gradual expansion of the stomach, eventually allowing a liter or more of food to enter into the cavity. This process is called *receptive relaxation*, since it involves the gradual relaxation of the rugae to accommodate the incoming food. This allows the stomach to easily expand its volume to about 1.06 quarts (1 liter), after which the tension of the stomach may cause discomfort. The volume at which this occurs varies with the individual, dietary habits, emotional state, and a number of other factors.

The peristaltic contraction of the smooth muscle that is responsible for mixing the food with gastric juice to produce chyme is initiated in the fundus. However, the areas of the stomach differ in the strength of the smooth muscle and thus the intensity of the peristaltic action. The contraction in the fundus is relatively weak, but becomes progressively stronger as it moves through the body and antrum. In the body of the stomach the contractions are not sufficiently powerful enough to provide a significant amount of mixing, allowing the continued digestion of carbohydrates in the bolus by salivary amylase. Thus in many regards the body of the stomach acts primarily as a storage site for incoming food.

As the contractions continue, the gastric juice mixes with the food to form chyme. This chyme is propelled downward into the narrower regions of the antrum, where the force of the peristaltic actions increases significantly. At the terminal end of the antrum is the pyloric sphincter, which serves to isolate the stomach from the small intestine. The pyloric sphincter is never completely closed, allowing for an almost continuous passage of water and other fluids into the duodenum. As the peristaltic contraction of the antrum approaches the phyloric sphincter, a small amount of chyme is moved through the opening and into the duodenum. However, the majority of the chyme is blocked from passing and is forced backed into the antrum for further mixing (see Figure 2.8).

Regulating Stomach Motility

The emptying of the stomach contents, also called *motility*, usually takes between two to four hours following completion of a meal and is depen-

TABLE 2.2. Factors Influencing the Movement of Chyme into the Duodenum

Stimulation	Inhibition
Distention of the stomach	Distention of the duodenum
Presence of partially digested proteins in stomach	Presence of fatty acids and carbohydrates in the duodenum
Gastrin	Cholecystokinin (CCK) Gastric inhibitory peptide (GIP)
Fluid chyme	Viscous chyme
Presence of alcohol or caffeine in stomach	

dent on a large number of factors. These factors either inhibit or stimulate the movement of the chyme and are summarized in Table 2.2. For the most part, actions of the stomach increase motility into the duodenum, while feedback from the duodenum inhibits movement of chyme through the pyloric sphincter. There are three distinct phases to stomach motility: the *cephalic phase*, *gastric phase*, and *intestinal phase*.

The prefix *ceph-* means "head," and the cephalic phase refers to the interaction of the brain with the stomach. If chemical receptors detect the smell or taste of food, a signal is sent to the medulla oblongata in the brainstem, which relays a signal along the **vagus nerves** to the submucosal plexus in the stomach. The submucosal plexus then stimulates the activity of chief and parietal cells, thus preparing the stomach for incoming food. A similar event occurs when a person thinks about food, especially those foods that that the individual enjoys. The emotional state of the individual, such as anger or anxiety, may inhibit these stimuli by activating the **sympathetic nervous system**. The sympathetic nervous system is involved with the *fight-or-flight response*, one aspect of which is to reduce activity in the gastrointestinal system so that blood may be redirected to muscles.

As its name implies, the gastric phase involves the activity of the stomach. Two factors influence its activity. First, the amount of distention, or stretching of the stomach lining, acts as an indicator of the fullness of the stomach. As the stomach fills, and the rugae relax, stretch receptors in the lining stimulate the release of gastrin by G cells in the mucosal lining of the antrum. Gastrin is released into the blood stream, where it returns to the stomach to stimulate the generation of gastric juice by the parietal and chief cells. The gastric juice has a normal pH of around 2.0; if this becomes more basic (or alkaline) then the secretion of gastrin is increased. (A reverse reaction occurs if the pH level of the stomach increases above 2.0.) The stretch receptors also stimulate the peristaltic contractions of the stomach.

As noted in the section on mechanical digestion, it is these contractions that are responsible for mixing the incoming food with gastric juice. As the contractions increase in strength, more of the chyme passes through the pyloric sphincter. The amount varies with the consistency, or fluidity, of the chyme. Under the ideal conditions, around 0.01–0.02 quarts (10–15 milliliters) of material may pass through the pyloric sphincter with each wave of contractions.

The duodenum of the small intestine may also regulate the activity of the stomach during the intestinal phase. Since the small intestine represents the major organ of digestion and absorption in the body (see Chapter 3), the duodenum must be ready to receive the incoming chyme for processing. The duodenum primarily has an inhibitory effect on stomach motility (see Table 2.2). Distention of the duodenum, due to the presence of a large volume of chyme, initiates a neural response called the *enterogastric reflex*, which through the action of the medulla oblongata decreases the strength of peristaltic contractions in the stomach, thus reducing the amount of chyme entering the duodenum. The presence of partially digestion carbohydrates and fats in the duodenum also activates an inhibitory pathway, but this pathway is based on the action of hormones. The action of salivary amylase, lingual lipase, and gastric lipase had previously started digestion of both carbohydrates and fats in the stomach. When these breakdown products reach the duodenum, they signal the release of *gastric inhibitory peptide* (GIP), *secretin*, and *cholecystokinin* (CCK) by the mucosa layer of the duodenum. These hormones are released directly into the bloodstream and influence the activity of a number of organs of the digestive tract (see Chapters 3 and 4), including the stomach. The secretions of the stomach are inhibited by secretin and GIP, while CCK and GIP reduce gastric motility. When this occurs fewer breakdown products are generated, less chyme enters the duodenum, and thus fewer hormones are produced.

When the gastric and intestinal regulatory mechanisms are combined, it allows a fine tuning of the gastrointestinal system to ensure that the optimal amount of material is being processed by the small intestine at all times.

Absorption of Nutrients

As mentioned previously, very few nutrients are absorbed through the lining of the stomach, primarily due to the presence of the mucus layer, which isolates the mucosa tissue from the hydrochloric acid. However, water and some ions are able to be absorbed directly into the circulatory system. In addition, both ethyl alcohol (the form found in alcoholic beverages) and acetylsalicylic acid (commonly known as aspirin) are able to penetrate the mucus layer and enter into the circulatory system. (The effects of these two compounds on the stomach will be discussed in more detail in Chapter 9.)

The Lower Gastrointestinal Tract: Small Intestine and Colon

The previous chapter described how the upper gastrointestinal (GI) tract, consisting of the stomach, esophagus and oral cavity, was involved with the processing of food for digestion. While there were some examples of enzymatic digestion in the upper gastrointestinal tract, the majority of the activity was associated with mechanical processing. The primary purpose of this mechanical digestion was to increase the surface area of the food so that the enzymes of the lower gastrointestinal tract can efficiently break down the nutrients into chemical forms that are able to be rapidly absorbed by the small intestine.

The lower gastrointestinal tract consists of two digestive organs called the small intestine and large intestine. Assisting with the operation of the small intestine are three accessory organs, the liver, gall bladder and pancreas (see Figure 3.1). The lower gastrointestinal tract serves two primary functions. First, the small intestine functions as the main organ of digestion and absorption in the human body. It is here that the bulk of nutrient processing is performed. The large intestine, or colon as it is commonly called, is primarily involved in the reabsorption of water and salts back into the body, and the preparation of the fecal material for excretion. This chapter covers the physiology of the small and large intestine and their interaction with the accessory organs of the digestive system. (A more detailed discussion of the accessory organs is presented in Chapter 4. Ailments of the lower gastrointestinal tract are detailed in Chapters 10 and 12.)

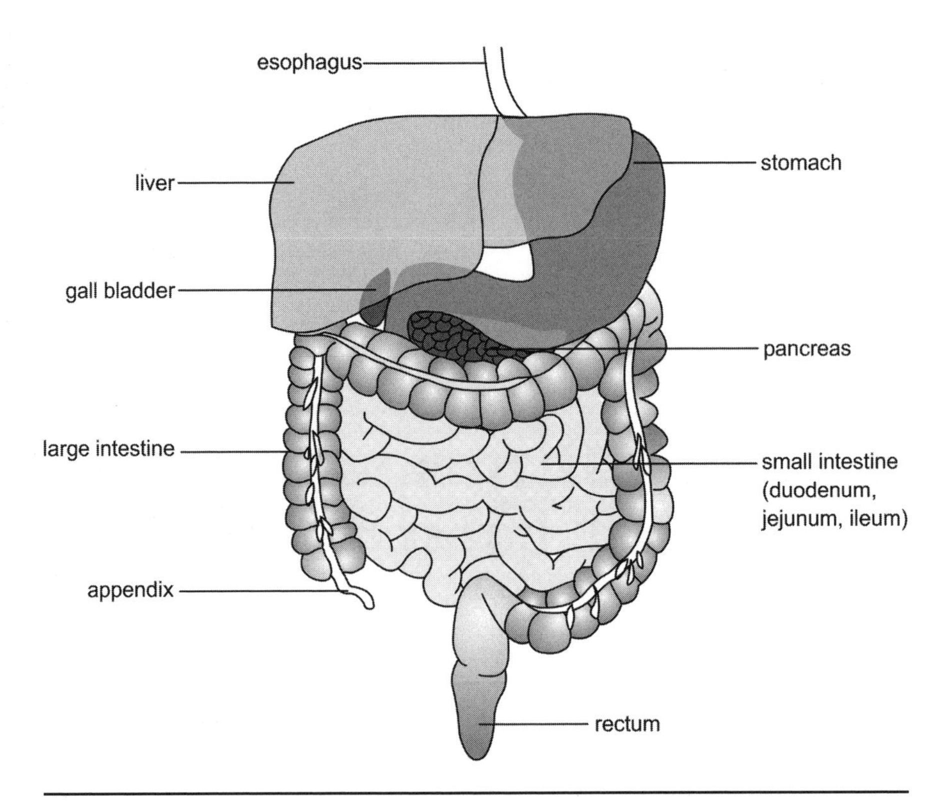

Figure 3.1. The lower gastrointestinal tract, giving the locations of the digestive organs and accessory glands.
The large intestine is frequently called the colon.

SMALL INTESTINE

The name of the small intestine is derived from its diameter, and not its overall size. The small intestine averages only approximately 1 inch (2.5 centimeters) in diameter, but in an adult can be over 10 feet (3 meters) in length. Although it may be of a small diameter, the small intestine represents the major site of digestion and absorption in the human body and is thus one of the more important organs of the digestive system.

The small intestine connects to the stomach at the pyloric sphincter and empties into the colon through the *ileocecal valve*, also called the ileocecal sphincter. For the purpose of study the small intestine is divided into three segments, although there is little difference in the physical appearance or structure of the regions. The first 7.8–9.8 inches (20–25 centimeters) of the small intestine, starting at the pyloric sphincter, is the duodenum. The next 2.7 yards (2.5 meters) of the small intestine is called the *jejunum* and the last section, about 2 yards (2 meters) in length and ending at the ileocecal valve, is the *ileum*.

The previous chapter examined the structure of the tissue layers in the gastrointestinal tract (see Chapter 2, Figure 2.7). As was the case with the stomach, the physical characteristics of these tissue layers vary in the lower gastrointestinal tract. The most significant of these differences occur in the mucosa and submucosa layers, the two innermost tissue layers. These layers interact directly with the interior cavity, or lumen, of the small intestine.

The most notable difference in the structure of the mucosa layer is the presence of numerous fingerlike projections called ***villi***. Unlike the folds in the mucosa layer of the stomach, which enabled it to expand in response to an incoming volume of food, the villi of the small intestine are involved in increasing the surface area to facilitate the absorption of nutrients. There may be as many as forty villi per square millimeter of mucosa, effectively increasing the surface area of the small intestine by a factor of ten. Within each villi (see Figure 3.2) are capillaries and portions of the lymphatic section called **lacteals**. As nutrients are absorbed into the villi (see the following section), they pass into either the capillaries or lacteals and are transported away from the small intestine. Along each of the villi are located four types of specialized cells. At the base of each villi, in pits called the intestinal glands (also known as the *crypt of Lieberkühn*), are the *Paneth cells*. These cells release lysozyme, an enzyme that protects the small intestine from bacteria. They also may move larger nutrient particles out of the lumen by the process of **phagocytosis**. Also located within the intestinal glands are *enteroendocrine cells*. As their name implies, these cells are actually part of the endocrine system and are responsible for releasing hormones such as gastric inhibitory peptide (GIP), secretin, and cholecystokin (CCK).

Further up the villi are located the *goblet cells*, which are responsible for secreting the protective mucus coating of the small intestine. While this mucus coating is not as thick as that found in the stomach, it serves to lubricate and protect the mucosa of the small intestine. Located along the length of the villi, but primarily above the region of the intestinal glands, is a layer of epithelial cells. These cells, also called *absorptive cells*, are the major site of nutrient absorption in the small intestine. These cells are unique in that the plasma membrane on

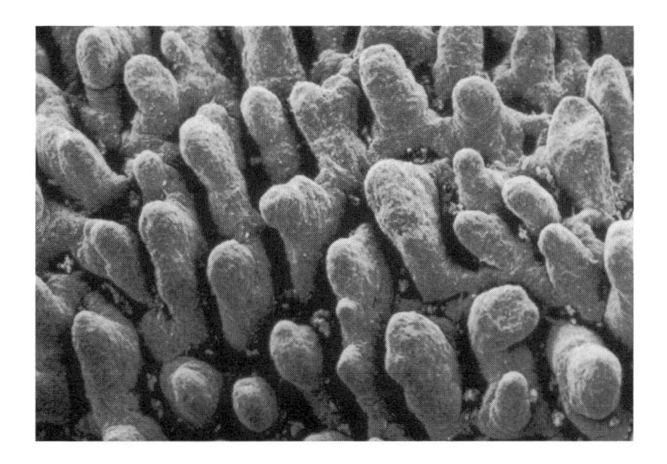

Illustration of bacteria in the small intestines. © Becker/Custom Medical Stock Photo.

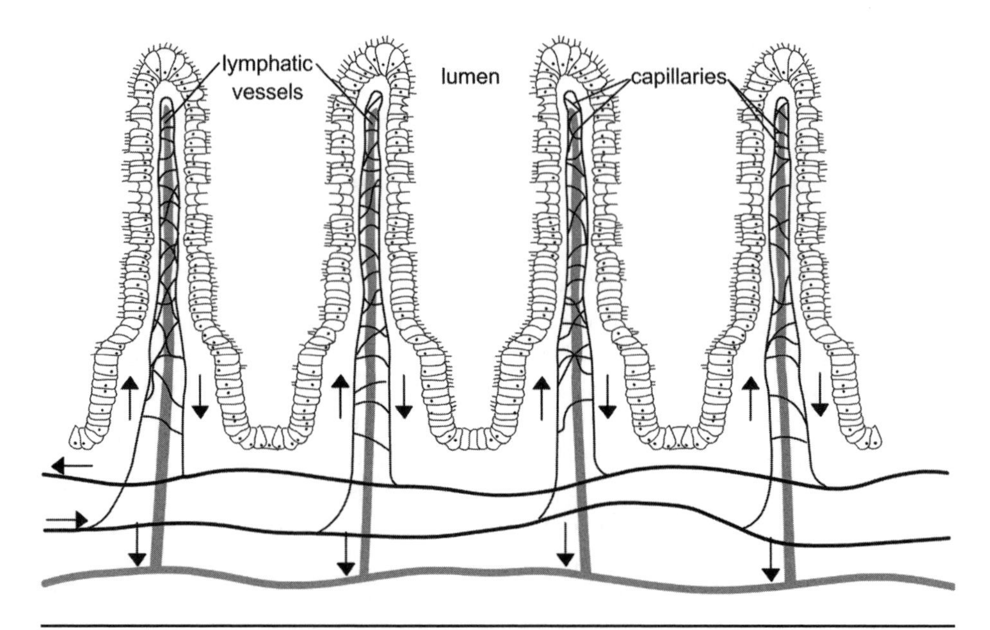

Figure 3.2. The structure of a villi of the small intestine.
Notice that each villi contains both capillaries and lymphatic vessels.

the lumen side of the cell contains a large number of small projections called *microvilli.* Each of these projections is about 1 micrometer (μm) long and a typical absorptive cell may have as many as 6,000 microvilli on its surface. A square millimeter of small intestine may contain up to 200 million microvilli. This increases the surface area of the small intestine by an additional factor of twenty. When combined, the villi and microvilli of the small intestine increase the overall surface area in this organ by 600 times, further disproving the idea that this is a "small" intestine. Under a microscope, the microvilli appear as a thin, fuzzy barrier on the lumen side of the absorptive cells. This is sometimes called the *brush border,* and it represents a region that is not only involved in nutrient absorption, but also in nutrient digestion by a group of enzymes called the brush border enzymes (see the following sections on the individual nutrient classes).

In the submucosa a group of specialized cells called the *duodenal glands,* also known as the *Brunner's glands,* release additional mucus into the lumen. This mucus is alkaline in pH, which helps to neutralize any hydrochloric acid from the stomach remaining in the food as it moves through the small intestine.

Movement of Nutrients

It takes approximately three to five hours for nutrients to transit the small intestine from the pyloric sphincter to the ileocecal valve. In the stomach the incoming food was mixed with gastric juice to form chyme. This chyme is moved along the length of the small intestine by two different types of contractions, *peristalsis* and *segmentation*. The peristaltic contractions in the small intestine are similar to those found in the stomach and esophagus. However, the contractions in the small intestine are much lower in intensity than those in the upper gastrointestinal organs.

The primary mechanism by which the chyme is moved through the small intestine is by segmentation. Unlike peristalsis, which is the rhythmic, sequential contraction of the smooth muscle layers of the gastrointestinal tract, the process of segmentation involves the localized contraction of small segments of the small intestine. These circular contractions squeeze the chyme against the mucosa layer of intestine, bringing the nutrients into direct contact with the microvilli of the absorptive cells. These contractions also serve to further mix the chyme with the secretions of the small intestine, gall bladder and pancreas (see following section). As the chyme is mixed by segmentation, it is slowly propelled along the length of the small intestine toward the ileocecal valve. Since the contractions of segmentation are not directional, as is the case with peristaltic contractions, some of the material will actually backup into the previous segment of the intestine. To ensure that the chyme has an overall movement toward the large intestine, the duodenum contracts more frequently (around twelve per minute) than either the jejunum or ileum (approximately nine per minute).

After the processing of a meal is complete, the small intestine enters into a "housekeeping" mode to remove the remnants of the chyme from the lumen. This consists of a series of weak peristaltic contractions that begin in the duodenum and contract for a short length of the intestine before ending weakly. The next contraction begins a little further down the intestine, and so on. The process is analogous to a sweeping action and is called the *migrating motility complex*. The entire process can take several hours to complete.

Although the small and large intestine are linked by similarities in their names, in reality the internal environments and physiology of these organs are vastly different. As will be described in the next section, the large intestine possesses a natural population of bacteria. If allowed into the small intestine, these bacteria could wreak havoc with the delicate tissues present there. The ileocecal valve is well designed to prevent the movement of materials into the small intestine from the large intestine. The folds of the valve are arranged to easily open as the chyme moving through the ileum of the small intestine exerts pressure against the it. However, if material in the large intestine presses against the valve, the pressure forces the folds tightly closed, preventing contamination of the small intestine. The valve is also under hormonal control. As food enters the stomach special cells (see Chapter 2) release a hormone called gastrin. Gastrin serves to relax the ileocecal valve, allowing for the emptying of the small intestine in response to an incoming meal.

Role of Accessory Glands

The digestive functions of the small intestine are assisted by the secretions of three accessory glands—the liver, gall bladder, and pancreas. This section will introduce the digestive functions roles of these glands. The detailed anatomy and physiology of these organs will be discussed in the next chapter.

One of the most important organs in the human body is the liver. The largest organ by weight, the liver coordinates activity between a number of body systems. The liver not only provides chemicals to assist in the digestive process, but it also filters and stores nutrients coming from the digestive system. (This aspect of liver physiology will be discussed in more detail in Chapter 4.) As an accessory organ to the small intestine, the liver provides a compound called *bile*, which assists in the process of lipid digestion. Bile is synthesized in the small intestine from a number of chemicals, including cholesterol, phospholipids, bile acids and water. Another ingredient is **bilirubin**, a waste product from the breakdown of worn-out red blood cells in the liver. Bilirubin does not play a role in digestion, but does give bile and the fecal material their color. Bile is continuously excreted into the *bile duct*, which connects the liver to the duodenum via the *hepatopancreatic ampulla*.

The gall bladder is a small sac, roughly pear-shaped, that is located just beneath the liver (see Figure 3.1). The sole purpose of the gall bladder is the storage of the bile salts from the liver between meals. Before entering the duodenum, the bile duct links up with the pancreatic duct to form the hepatopancreatic ampulla. At the duodenum end of this structure is a small valve called the *Sphincter of Oddi*. This sphincter is normally open when chyme is present in the duodenum, but closes between meals. Since bile production by the liver is a continuous event, the bile leaving the liver backs up and enters the gall bladder to be stored. As food enters the duodenum, CCK is released from the enteroendocrine cells in the mucosa of the small intestine. This hormone signals the gall bladder to contract, releasing its contents, as well as acting as a signal for the Sphincter of Oddi to relax. (The recycling of bile salts will be discussed in more detail in Chapter 4; ailments of the gall bladder will be covered in Chapters 11 and 12.)

The third major accessory organ is the pancreas. It is located just below the stomach, adjacent to the small intestine (see Figure 3.1). Like the liver, the pancreas performs a variety of functions in the human body. As an accessory organ for the digestive system, the pancreas is responsible for providing the majority of the digestive enzymes needed for nutrient processing in the small intestine. The cells of the pancreas (see Chapter 4) produce a

TABLE 3.1. The Sources of the Digestive Compounds of the Small Intestine

Target Nutrient	Compound	Source
Carbohydrates	Pancreatic amylase	Pancreas
Lipids	Bile	Liver
	Pancreatic lipase	Pancreas
Proteins	Chymotrypsin	Pancreas
	Trypsin	Pancreas
	Carboxypeptidase	Pancreas
	Elastase	Pancreas
Nucleic Acids	Deoxyribonuclease	Pancreas
	Ribonuclease	Pancreas

colorless liquid called *pancreatic juice*. This colorless, watery mixture is collected into the pancreatic duct, which later joins the bile duct to form the hepatopancreatic ampulla, which empties into the duodenum. The major enzymes present in the pancreatic juice are listed in Table 3.1. In addition to these enzymes, pancreatic juice contains a compound called *sodium biocarbonate*. This compound is slightly alkaline and serves to neutralize the acid in the chyme from the stomach. This establishes the correct pH for optimal enzyme activity in the small intestine.

Overview of Nutrient Processing

As mentioned previously, the small intestine is the primary organ of digestion and absorption in the human body. In general, the small intestine has to process two very broad classes of nutrients. The **hydrophilic** ("water-loving") molecules, which include the monosaccharides and many of the amino acids, are easily transported, digested and absorbed by the small intestine. The second class is the hydrophobic ("water-fearing") molecules, of which fats and cholesterol are the major examples. Due to their chemical properties, hydrophobic molecules will require more elaborate processing and transportation systems.

For each of the four major classes of organic nutrients, the carbohydrates, lipids, proteins, and nucleic acids, the job of the small intestine is to first breakdown the large complex structures of these nutrients into units that are small enough to be transported into the villi. (There are some exceptions, but for most nutrients these are the monomers discussed in Chapter 1.) The following sections will detail the digestion and absorption of each of the major classes of nutrients. It is important to remember that, for the most part, the processing of nutrients occurs simultaneously throughout the length of the small intestine.

CARBOHYDRATE DIGESTION AND ABSORPTION

Carbohydrate digestion began in the oral cavity with the activity of the salivary amylase enzyme. While this enzyme only acts on the incoming food for a brief period of time before becoming inactivated by the hydrochloric acid of the stomach, it does initiate the breakdown of starches and other polysaccharides into disaccharides and monosaccharides. The majority of the polysaccharide digestion is conducted in the small intestine by a secretion of the pancreas called *pancreatic amylase*. Pancreatic amylase is active in the lumen of the small intestine. Like salivary amylase, this enzyme can break down starch and glycogen, but not plant polysaccharides such as cellulose, commonly called fiber (see "Soluble versus Insoluble Fibers"). These remain undigested and unprocessed until they reach the large intestine and provide much of the bulk of the food.

Soluble versus Insoluble Fibers

The terms *fiber* and *cellulose* are frequently collectively used to describe the class of complex carbohydrates that are undigestable by the human body. While there are many different types of fibers, they can be grouped into two general classes: the soluble and insoluble fibers. In general the soluble fibers slow down movement of material through the GI tract and delay the absorption of glucose into the villi. In comparison, the insoluble fibers may speed the transit of material through the GI tract as well as providing bulk to the fecal material. Like the soluble fibers, their presence also slows the absorption of glucose. The major forms of soluble and insoluble fibers are listed below.

- *Cellulose.* This is the most common fiber and is found in almost all plant material. It is the primary component of the cell wall of plants and is insoluble in water.
- *Lignin.* This is another insoluble fiber that is common in the tough parts of woody plants and some seeds.
- *Hemicellulose.* Cereal bran is the prime source of this fiber. In most cases it represents an insoluble fiber, but some sources are considered to be soluble.
- *Pectins.* A soluble fiber that is common in fruits, especially apples and citrus fruits.
- *Gums.* Another soluble fiber, which is sometimes also called mucilage.

To be absorbed into the villi, carbohydrates must be in their simplest form, as monosaccharides. In most cases, the action of the salivary and pancreatic amylases generates a compound called *dextrin*, a short chain of glucose molecules. The final breakdown of the carbohydrates into monosaccharides is performed by a group of enzymes physically embedded in the brush border of the small intestine. These enzymes are named by the specific carbohydrate substrate that they recognize. For example, the *dextrinase* enzyme breaks down the short-chains of glucose (dextrins) into single glucose units. The *maltase* enzyme digests the disaccharide sugar maltose into two glucose units. Sucrose, another of the disaccharides that consists of glucose and fructose, is digested by the *sucrase* enzyme and lactose, a sugar commonly found in milk, is broken into the monosaccharides galactose and glucose by the action of the *lactase* enzyme. (A condition called lactose intolerance, in which inadequate amounts of lactase are produced, will be discussed in Chapter 10.)

Once the carbohydrates in the chyme have been digested into one of the three monosaccharides (fructose, glucose, or galactose), they are ready for absorption into the villi of the small intestine (see Figure 3.3). The molecules themselves are too large to pass directly through the membranes of the intestinal cells. Fructose is absorbed by a process called **facilitated diffusion**, in which a protein channel in the membrane of the epithelial cells aids in the movement of the fructose into the villi. The movement of glucose and galactose is slightly different. To move these molecules into the epithelial cell, the cell couples their movement with the transport of sodium ions (Na^+) into the cell. This is a form of **active transport**, and as such requires energy.

Once these monosaccharides are in the cytoplasm of the epithelial cell, they are rapidly moved into the capillaries of the villi by the process of facilitated diffusion. There they enter the circulatory system and proceed to the liver for additional processing (see Chapter 4). The rapid movement of these molecules out of the epithelial cells ensures that the interior of the cell has a low concentration of the monosaccharides, thus aiding in the absorption of sugars from the lumen.

PROTEINS

The digestion of proteins was initiated in the stomach with the activity of the enzyme pepsin. Recall that pepsin is a general enzyme, which primarily serves to disrupt the complex three-dimensional structure of the proteins in the chyme. Thus when the proteins reach the small intestine, they are primarily in the form of small chains called peptides. In general, protease enzymes are powerful catalysts and thus are secreted in an inactive form until needed by the body. This prevents the unwanted digestion of the mucosal lining. For example, in the stomach the pepsin enzyme was gen-

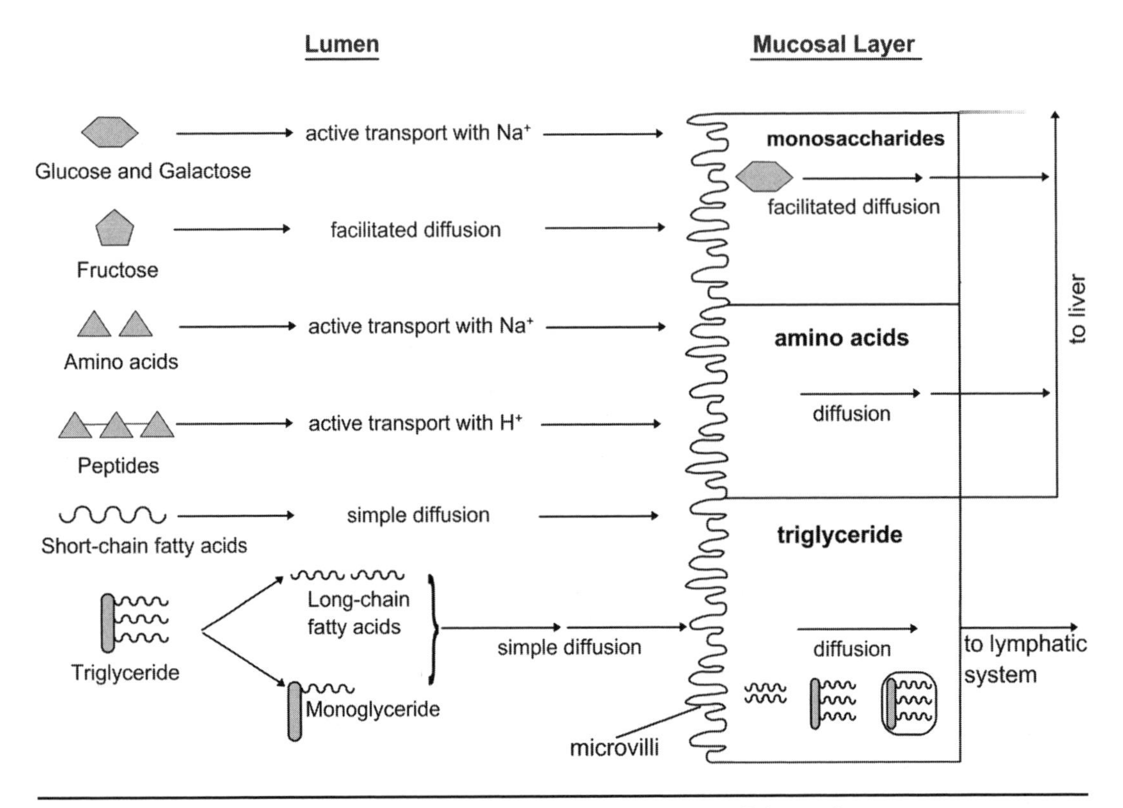

Figure 3.3. An overview of nutrient absorption in the small intestine.
Note how monosaccharides and most amino acids proceed directly to the liver, while the triglycerides enter in the lymphatic system. Triglycerides are first disassembled in the lumen and then reassembled into chylomicrons for transport in the lymphatic system.

erated from a molecule called pepsinogen under the correct pH conditions. A similar event occurs in the small intestine. The pancreas releases a **proto-enzyme** called *trypsinogen* into the lumen of the duodenum via the pancreatic duct. Trypsinogen is inactive and cannot begin the breakdown of proteins until it is first activated by *enterokinase,* an enzyme that is present in the brush border region of the villi. The enterokinase cleaves off a small portion of the trypsinogen molecule, forming the enzyme *trypsin.* In turn, trypsin activates three reactions with chymotrypsinogen, proelastase, and procarboxypeptidase to produce *chymotrypsin*, *elastase* and *caboxypeptidase*, respectively. These molecules are all proteases that are active in the lumen of the small intestine. While each targets a different structural part of the protein, their overall function is to digest the proteins into smaller peptides, at times releasing individual amino acids.

Present in the brush border region are two other enzymes. The first are the *aminopeptidases*, which are responsible for removing the terminal amino acid from the end of the peptide chain. The second is an enzyme called *dipeptidase*, which breaks the remaining peptide bond holding two amino acids together. The end result of this activity is free amino acids, which may now be absorbed into the villi.

The absorption of the individual amino acids is very similar to the process outlined previously for carbohydrates. In most cases, the movement of amino acids across the membrane is an active process that is coupled to the movement of sodium ions (Na^+), although in some cases the process is coupled to the transport of hydrogen ions (H^+). Once in the epithelial cells, the amino acids diffuse into the capillaries of the villi, to be transported by the **hepatic portal system** (see Chapter 4) to the liver for processing.

There are two additional aspects of protein digestion to be mentioned. While previously it was presented that the three regions of the small intestine are fundamentally the same, there are some minor differences with regards to protein digestion and absorption. The majority of protein digestion occurs in the duodenum and jejunum, with only minimal activity in the ileum. Also, any protein that enters the small intestine is subject to the digestive process outlined above. This includes not only proteins from food sources, but also the proteins found in worn-out mucosal cells, enzymes such as pepsin and salivary amylase, bacterial proteins, and miscellaneous proteins such as bilirubin, that are excreted from the liver and pancreas.

LIPIDS

The processing of lipids by the small intestine differs significantly than that described previously for carbohydrates and proteins. This is primarily due to the fact that lipids are hydrophobic molecules, and as such are not easy to work within the hydrophilic environment of the lumen of the small intestine.

Prior to the entry of lipids into the small intestine there was a small amount of processing performed by the lingual lipase in the oral cavity. This enzyme primarily serves to initiate triglyceride digestion, but is only active for a relatively short period of time before entering the stomach. In the stomach, gastric lipase may act on short-chain fatty acids, such as those found in milk products. However, this enzyme does not make a significant contribution to lipid processing. Thus, the real first level of lipid digestion and absorption occurs in the small intestine.

As the chyme passes the pyloric sphincter it is mixed with secretions of the liver, gall bladder and pancreas. Recall from the previous section that the liver produces bile salts, which may be temporarily stored and concentrated within the gall bladder. Bile emulsifies, or breaks down, droplets of fats in the chyme into smaller particles. This is not enzymatic digestion, since the individual lipid molecules are not the target, but rather the interaction between the fat molecules. The result is small droplets of lipids with a diameter of about 0.039 inches (1 millimeter). This drastically increases the surface area for digestion, speeding the overall processing of the lipids.

The enzyme responsible for the digestion of lipids is pancreatic lipase. Pancreatic lipase enters the small intestine through the duct called the *hepatopancreatic ampulla*, along with bile from the gall bladder and liver. This lipase breaks down triglycerides into small fatty acids chains and monoglycerides, which consist of a single fatty acid chain connected to the glycerol backbone (see Chapter 1 and Figure 3.3). In these forms the molecules can then be absorbed directly into the epithelial cells of the villi by the process of diffusion.

However, most of the triglycerides that enter into the small intestine contain long-chain fatty acids, which due to their size can't diffuse into the villi. For the long chain fatty acids, and similar hydrophobic molecules such as cholesterol, a different process exists to move the nutrients out of the lumen and into the body. When combined with the bile salts released from the liver, the lipids and cholesterol form spheres called *micelles*. Each micelle consists of an outer shell of approximately thirty to fifty bile salt molecules. Micelles are **amphipathic** molecules, meaning that they have both polar and nonpolar regions, enabling them to interact with both hydrophobic and hydrophilic molecules. The hydrophobic lipids are carried within the center of the micelle. When the micelle reaches the cell membrane of the epithelial cells in the villi, the lipids and other hydrophobic molecules in the core of the micelle are able to diffuse across the membrane. The sphere of bile salts is then able to return to the lumen to pick up more hydrophobic lipids. In other words, the micelle acts as a shuttle by providing a hospitable environment for the movement of large hydrophobic molecules.

Since bile salts represent a reusable resource for the digestive system, they are recycled in the small intestine. The bile salts that were initially released in the duodenum are reabsorbed in the ileum of the small intestine. There they enter into the portal circulatory system (see Chapter 4) and are returned to the liver. This circular recycling of bile salts is sometimes called the *enterohepatic* circulation.

One more important aspect of lipid processing occurs in the small intestine. With hydrophobic nutrients, such as sugars, the nutrients that are absorbed by the small intestine quickly diffuse into the capillaries of the villi, where they then enter the circulatory system. (The movement of nutrients in the circulatory system will be discussed in greater detail in Chapter 4.) However, lipids by nature do not interact well with an aqueous environment, and their large size would quickly clog the narrow capillaries contained within the villi. Instead, lipids are packaged into a special form of lipoprotein called a *chylomicron*. Chylomicrons are protein-covered balls of lipids, cholesterol and phospholipids (see Figure 3.3). The role of the chylomicron is to move the lipids into the lacteal of the villi, where it then enters into the lymphatic system. (The movement of these nutrients by the lymphatic system will be covered in more detail in Chapter 4.)

NUCLEIC ACIDS

Recall that nucleic acids represent the genetic material of living organisms, and thus are present in most of the material being processed by the small intestine. Since DNA and RNA (ribonucleic acid) both consist of long chains of nucleotides, they are digested in a similar manner by the small intestine. The fact that the DNA is double-stranded, and typically a longer polymer, has little influence on the properties of nucleic acid digestion.

Along with its previously mentioned enzymes, the pancreas secretes pancreatic ribonuclease and deoxyribonuclease, which act on RNA and DNA respectively. The purpose of these enzymes is to cleave individual nucleotides from the polymer. In many ways these enzymes function similarly to the pancreatic proteases mentioned earlier. Once a nucleotide is removed from the polymer, it is further digested by brush border enzymes called *nucleosidases* and *phosphatases*. These enzymes break the nucleotide down into its constituent sugars, phosphates, and nitrogenous bases for absorption into the villi by active transport. Once there they move into the capillaries by diffusion and enter the portal circulatory system (see Chapter 4).

A woman drinking water out of a glass. © Thinkstock.

WATER

While water is an integral part of the digestive tract, and one of the more important nutrients in the human body, it is not "digested" in the same manner as the organic nutrients just mentioned. Instead, it is absorbed by the villi of the small intestine into the circulatory system, and to a lesser extent the lymphatic system. Most physiologists believe that water moves from the lumen of the intestine into the epithelial cells by the process of **osmosis**, or the diffusion of water. This passive process is dependent on the concentration of solutes and has long been recognized as the prime mechanism of water movement by biological systems. However, researchers have begun to discover that many organisms possess specialized channel proteins, called *aquaporins*, that allow for the rapid movement of water across a plasma membrane. Whether aquaporins are present in human epithelial cells, and what their contribution to the overall movement of water is, remains to be determined. Regardless, the small intestine processes a tremendous volume of water daily. Almost 10 quarts (close to 9.3 liters) of water enter the small intestine daily, most of it (7.4 quarts; approximately 7.0 liters) comes from the secretions of the accessory glands (4.2 quarts; about 4 liters), stomach (2.1 quarts; 2 liters), and small intestine (1.06 quarts; 1 liter). The remainder (2.4 quarts; an average of 2.3 liters) is obtained from the ingested food and liquids. The small intestine reabsorbs almost 90 percent of this volume, with the remainder passing into the large intestine. Ailments of the small intestine, the emotional state of the person, and environmental factors, can all easily interfere with this process, and will be discussed in more detail in Chapter 10.

VITAMINS

Vitamins are frequently assigned to two general classes, those that are water soluble (vitamin C and the B-complex vitamins) and those that are fat-soluble (vitamins A, D, E, and K). These general classes also apply to the approach that the small intestine takes in absorbing these important compounds. The water-soluble vitamins are treated in much the same manner as monosaccharides and amino acids, meaning that they are actively transported into the epithelial cells and then move by diffusion into the capillary of the villi. Fat-soluble vitamins may either move into the epithelial cells by diffusion, or through the action of the micelles. They are typically then loaded into chylomicrons for transport into the lymphatic system.

There are some vitamins that have special processing in the small intestine. Vitamin B_{12}, sometimes also called *cobalamin*, is typically found in protein rich foods. The pH of the stomach releases the vitamin, which then binds with an intrinsic factor before entering the small intestine. In the small intestine vitamin B_{12} is absorbed into the epithelial cells, where it then returns to the liver via the enterohepatic circulation. The liver continuously secretes both vitamin B_{12} and folate into the bile. Since folate is associated with the health of rapidly dividing cells, and vitamin B_{12} is needed to activate folate, this mechanism ensures that the rapidly dividing epithelial cells of the small intestine are provided with a source of these important vitamins.

MINERALS

Minerals, like water, are not organic nutrients and thus are not digested by the small intestine. However, the small intestine does represent an important location of absorption for many of the minerals in a human diet.

Unlike the other nutrient classes, mineral absorption in the small intestine is not guaranteed. Many foods contain chemicals that actively bind nutrients, reducing their ability to be absorbed. Thus for minerals it is often more correct to refer to their **bioavailability**, and not necessarily the total quantity in the food. Examples of these binders are oxalic acid, found in leafy vegetables such as spinach, and phylic acid, a compound frequently found in grains and beans (legumes).

There are many different minerals, each with its own unique absorption properties. For nutritional purposes, minerals are classified as being either trace or major, depending on the quantity that is required in the diet (see Chapter 1). However, since minerals are not digested enzymatically, as was the case with the organic nutrients, the activity of the small intestine is confined to absorption only. Most of the minerals behave as hydrophilic molecules (such as potassium and sodium), but a few display hydrophobic characteristics. In most cases, minerals are absorbed by active transport into the villi.

Two minerals of special interest in examining the physiology of the small intestine are calcium and iron. Calcium is an important nutrient for muscle contraction, and has the secondary function of providing strength to bone (see the Muscular System and Skeletal System volumes of this series for more information). Calcium is usually brought into the digestive system in the form of a salt, and is kept in a soluble form by the acidic nature of the stomach. The efficiency of the small intestine in absorbing calcium is based upon a number of factors. In general, an adult human is able to absorb about 30 percent of the calcium found in food, although this value may vary depending on age, sex, gastrointestinal tract health, and emotional state. For example, young children frequently absorb up to 60 percent of ingested calcium and the value in pregnant women can reach 50 percent. (Other factors that influence calcium absorption are listed in Table 3.2.) However, the greatest factor that influences the absorption of calcium is the presence of vitamin D. Vitamin D, a fat-soluble vitamin, actually functions as a hormone, in that is it manufactured by one organ of the body to influence the activity of a second organ. (The discovery of the importance of vitamin D in the body is covered in Chapter 7.) Vitamin D may also be found in some foods, such as milk products, where it is frequently added to enhance calcium absorption. Once activated, vitamin D stimulates the small intestine to produce a calcium binding protein, which in turn facilitates the movement of calcium into the villi.

The absorption of iron is slightly more complex. As was the case with calcium, all of the iron that is ingested is not absorbed. In fact, as little as 10 percent of the available iron is absorbed in an adult male, and only 15 percent in an adult female. Children have slightly higher percentages (up to 35 percent) but the reality is that most of the iron that is present in food is excreted with the feces. In addition, iron may exist in one of two ion

TABLE 3.2. Factors Influencing Calcium Absorption in the Small Intestine

Inhibitory Factors	Enhancing Factors
Presence of oxalates and phylates	Presence of growth hormones
High fiber diets	Presence of lactose in the chyme
High phosphorus intake	Equal concentrations of phosphorus
	Vitamin D

forms: ferrous iron (Fe^{2+}) or ferric iron (Fe^{3+}). Of these, ferrous iron is more easily absorbed. The source of the iron also plays a role in iron absorption. Iron that is present in animal flesh, called *heme* iron, is more easily absorbed that iron that originates in plant material (non-heme iron).

As was the case with calcium, several environmental factors contribute to the absorption of iron from the chyme. Recently it has been discovered that vitamin C, a water-soluble antioxidant vitamin, has the ability to keep iron in its ferrous form, thus increasing its bioavailability. In addition, many meat products contain a substance called *MFP factor* that increases iron absorption. However, the iron processing by the small intestine can also be inhibited by the presence of a number of compounds (see Table 3.3).

One additional interesting feature of iron physiology is the ability of the small intestine to act as temporary storage site. Unlike most nutrients, which quickly move through the absorptive cells of the villi into the circulatory or lymphatic systems, the mucosal layer may actually store iron using a special protein called *mucosal ferratin*. This protein binds iron and releases it to *mucosal transferrin* when needed by the body. Transferrins are iron-transport proteins. Mucosal transferrin then transfers the iron to *blood transferrin* (sometimes just called transferrin) for movement into the body.

LARGE INTESTINE

The name large intestine is derived from its diameter (2.5 inches; 6.5 centimeters), not its length (1.37 yards; 1.5 meters). The large intestine begins at the *ileocecal valve*, which serves as the boundary between the small and

TABLE 3.3. Factors Influencing Iron Absorption in the Small Intestine

Inhibitory Factors	Enhancing Factors
Presence of phylates	Citric acid and lactic acid
High fiber diets	Some sugars
Presence of phosphorus and calcium	Vitamin C
Food additives such as EDTA (ethylenediamene tetra acetate)	MFP factor

large intestines. The terminal portion of the large intestine, and the entire gastrointestinal tract, is the *anus*. The large intestine is vastly shorter than the small intestine and differs significantly both in anatomy and function. The large intestine is often mistakenly considered the location in the body where waste material is generated. In reality, this organ is more of a recycling center and temporary storage location than a waste disposal site. The large intestine is comprised of three distinct regions: the *cecum*, the colon, and the *rectum* (see Figure 3.4). The colon is subdivided into four zones, called the *ascending colon, transverse colon*, *descending colon*, and *sigmoid colon*, based on their orientation and position in the body cavity. The physiology of these zones is fundamentally the same, and the names are used for descriptive purposes only. Because the majority of the large intestine is comprised of the colon, the term *colon* is frequently used as a common name for the large intestine.

While the appendix has historically been considered a part of the large

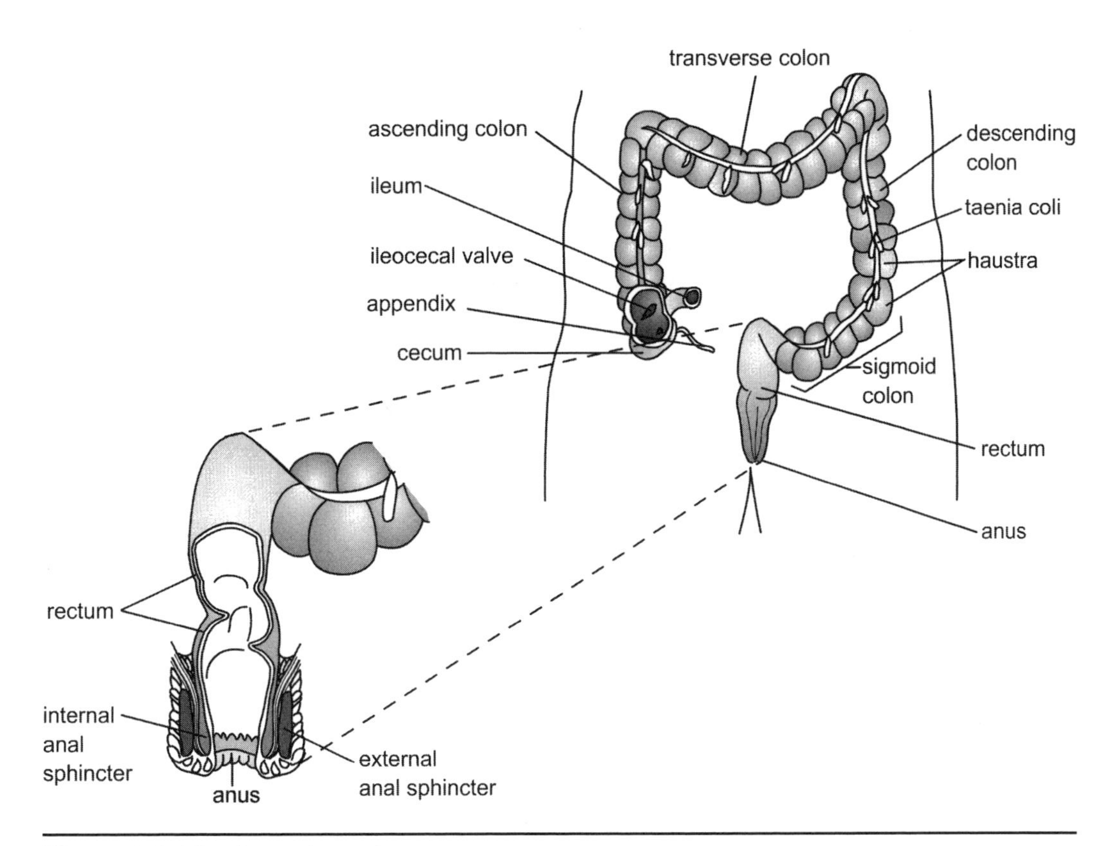

Figure 3.4. The large intestine.
Commonly called the colon, the large intestine consists of three major sections, named for their orientation in the abdominal cavity. The large intestine terminates in the rectum.

intestine, and often thought of as a **vestigial** organ, it is actually made from the same type of tissue as the lymph nodes and thus is now considered to be part of the lymphatic and immune systems. **Lymphocytes** housed in this area help protect the digestive system from pathogenic microorganisms and thus the appendix serves as an important first line of defense for the lower GI tract. However, ailments of the appendix are frequently caused by problems with the digestive system (see Chapter 10).

Movement of Material

The large intestine receives approximately 15 fluid ounces (500 milliliters) of material daily from the small intestine. Under normal conditions, the majority of this material is the undigested remnants from the small intestine (see following section). The colon regions of the large intestine are structured differently than the small intestine. Whereas the small intestine utilized circular and longitudinal patterns of smooth muscle to power the contractions necessary to move the food, the colon instead possesses three bands of smooth muscle. Rather than surrounding the GI tract, these muscles are arranged to run the length of the intestine, called the *taeniae coli*. The arrangement of the taeniae coli causes the exterior surface of the large intestine to resemble a series of small pouches, called *haustra* (see Figure 3.4). The haustra are not permanent structures, but actually change position slightly based upon the contractions of the taeniae coli.

The movement of the material through the large intestine is a much slower process than in the small intestine. This gives ample time for the sections of the colon to reabsorb important nutrients such as water. In the small intestine, the contractions of the smooth muscle (segmentation) occurred at a rate upwards of twelve times a minute. In the large intestine, the contractions may occur several times an hour. These contractions are called *haustral contractions*, and the entire process is called *haustration*. Also unlike the small intestine, in which the contractions were controlled to move the chyme through the length of the intestine, in the colon the haustral contractions are more regional. This causes the material to move back and forth between haustra, further increasing the time that the material is in the system.

There are several conditions that may cause a synchronization of these haustral contractions, resulting in a uniform movement of material toward the rectum and anus. The first of these is called a *mass movement*. Mass movements occur several times daily, usually following meals, and are characterized as synchronous contractions of the first two sections of the colon (ascending and traverse). This contraction propels the food into the descending colon, moving the material there into the sigmoid colon and rectum. Although a mass movement can occur without food entering the stomach, the body does possess a mechanism to clear the intestines to prepare for incoming food. As food enters the stomach, it triggers the *gastro-colonic reflex*. This reflex action causes contractions along the entire length of the intestines, moving food from the small intestine into the colon, and driving the undigested material in the colon into the rectum. The gastro-colonic reflex is often accompanied by an urge to defecate.

The final movement of material out of the gastrointestinal tract is called defecation and the factors that cause it to occur are called the *defecation reflex*. Defecation is actually a complex process since it involves both voluntary and involuntary actions of the anus. The anus, the terminal sphincter of the gastrointestinal tract, consists of both smooth and skeletal muscle. The internal anal sphincter (see Figure 3.4) is comprised of smooth muscle and thus is under involuntary control. During the defecation reflex, the smooth muscle of this sphincter relaxes. At the same time, the rectum and sigmoid colon contract, moving the contents toward the external anal sphincter. This sphincter is made of skeletal muscle, and thus is under voluntary control. If the external anal sphincter is relaxed, defecation occurs. If not, then the urge can be controlled, although this may result in excess water being absorbed from the feces, causing constipation (see Chapter 10). Although the external anal sphincter may be closed, it is still possible to force intestinal gas (*flatus*) out through a narrow opening in the anus, thus partially relieving pressure in the rectum.

Digestion and Absorption

Of the quarts of material that enter the digestive tract each day, only about 0.53 quarts (0.5 liters) actually ends up in the large intestine. The small intestine is highly effective as an organ of digestion and absorption, thus the material reaching the large intestine usually contains only undigested material, such as cellulose, some water, salts, and bilirubin from the liver. Since the amount of usable nutrients is severely limited by the time the food material reaches the large intestine, there is no enzymatic digestion conducted by the cells of the large intestine. While no digestive enzymes are secreted by the large intestine, the mucosal cells along the length of the organ secrete an alkaline mucus which serves to lubricate the internal lining of the large intestine, and protect it from any acids produced by fermenting bacteria in this region. There is some minor breakdown of the bilirubin from the liver, and this accounts for the characteristic color of the fecal material.

The large intestine also lacks the complex internal structure found in the small intestine. Instead of a network of villi and microvilli, the interior surface of the large intestine is for the most part smooth. This reduced surface area limits the ability of the large intestine to be a major organ of absorption. However, the decreased surface area is slightly compensated for by slowing the movement of material through the organ. The relative lack of segmentation and peristaltic contractions in the large intestine means that the material is present in the intestine for a longer period of time, allowing for more (although slower) absorption of selected nutrients. The action of the haustral contractions also serves to move the material back and forth within the colon, further slowing the movement of material. Dietary factors, namely the amount and types of fiber (see Chapter 1) also influence the rate of movement. In addition, health factors such as age, stress, and disease, all contribute to the speed at which the material transits the large intestine (see Chapter 10). Of the 15 fluid ounces (500 milliliters) of the material that enters the large intestine, about 10.5 fluid ounces (350 milliliters) is reabsorbed, with the remaining volume exiting through the anus as feces.

The colon primarily absorbs water and salts, although it may also take in other nutrients, such as glucose and vitamins, that may be present (see following section). Salts in the colon normally consists of both sodium and chloride ions, both of which are essential nutrients, and are reabsorbed. In addition, water is reabsorbed by osmosis, but a significant amount remains in the feces to lubricate it. The final daily fecal volume of 4.5 fluid ounces (150 milliliters) usually is two-thirds water and one-third solid material. Most of the solids is actually bacterial mass, with bilirubin from the liver and cellulose accounting for the remainder.

The slower movement of material also gives microorganisms, such as bacteria, the opportunity to establish populations. However, unlike the remainder of the gastrointestinal tract, where the presence of bacteria causes problems (see Chapter 10), the colon actually contains a natural **flora** of bacteria that make a positive contribution to human physiology. The bacteria of the large intestine are in a **symbiotic** relationship with their human host. The human host provides a stable environment, with ample water and nutrients. In return, bacteria such as *Escherichia coli* and *Clostridium* species break down undigested fiber to glucose. Some of this glucose may be absorbed by the colon. In addition, some B vitamins and vitamin K, which is involved in the clotting response of the circulatory system, are also produced. More importantly, the presence of large colonies of beneficial bacteria inhibits the growth of pathogenic bacteria that may have entered through the ileocecal valve. By some estimates there may be as many as 500 different species of bacteria in the human colon, with natural variations in strains and species between individuals and populations.

The Accessory Organs

The previous two chapters have examined the movement of food through the gastrointestinal system. As noted previously, there are two types of organs in the digestive system. Digestive organs, such as the stomach and small intestine, form the conduit through which food is moved and processed in the body. Associated with the operation of these organs are the accessory organs. In general, the accessory glands contribute important lubricants, enzymes, and chemicals that are required for the operation of the digestive organs. There are four accessory organs: the salivary glands, liver, gall bladder and pancreas (see Table 4.1). Figure 4.1 gives the position of these organs in the human body. Both the liver and pancreas have additional functions in the body. These functions will be discussed in other volumes of this series.

SALIVARY GLANDS

The salivary glands consist of three major pairs of glands that are located within the oral cavity (Figure 4.2). As previously mentioned in Chapter 2, these are called the parotid, sublingual, and mandibular glands. There are also minor salivary glands, called the *buccal* glands, located in the linings of the cheek. Located under the tongue, along the base of the mouth, are the sublingual glands. Excretions from these glands are moved into the oral cavity by a short duct called the lesser sublingual duct. Located just below the sublingual glands, along the jawbone (mandibula), is the submandibular gland, which is connected to the oral cavity by the submandibular duct. Just in front of each ear are the parotid glands. These are connected to the oral

TABLE 4.1. Summary of Accessory Gland Contributions to Digestion

Organ	Role in Digestion
Salivary Glands	Provides salivary amylase for CHO digestion Provides mucus to lubricate oral cavity and esophagus
Liver	Manufactures bile for lipid digestion Central role in carbohydrate, fat, and protein metabolism
Gall Bladder	Stores bile
Pancreas	Manufactures sodium bicarbonate Manufactures pancreatic digestive enzymes

cavity via the parotid duct, which enters close to the molars in the rear of the mouth.

The salivary glands play several important roles in the overall physiology of the oral cavity. First, they actively moisten the oral cavity, which greatly aids not only in the swallowing of food, but also in the process of

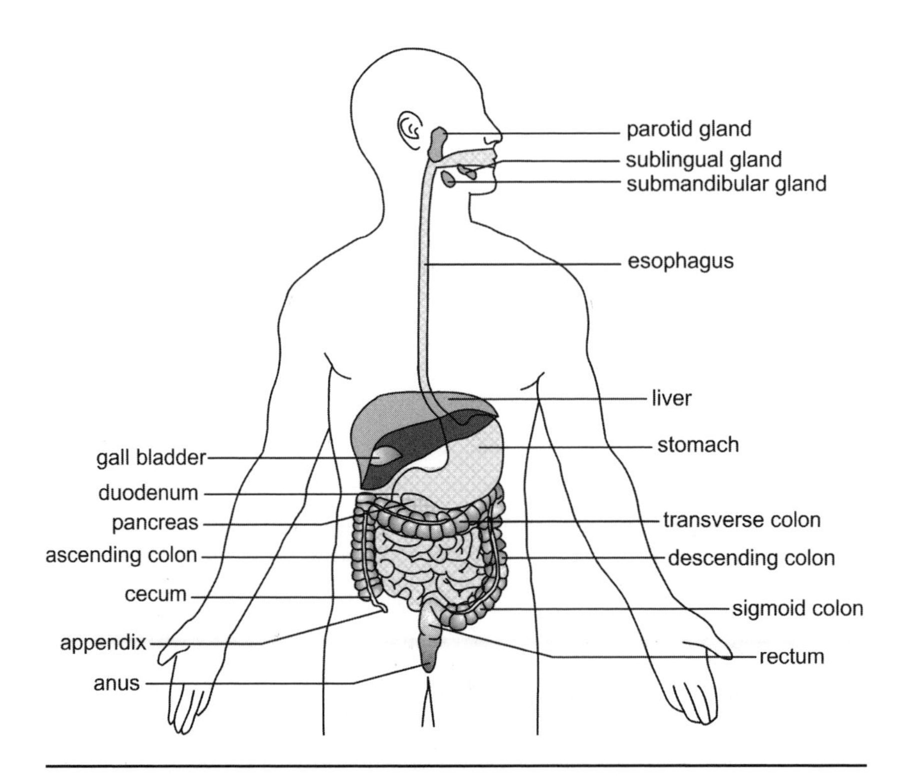

Figure 4.1. The gastrointestinal system.
This diagram shows the location of the four accessory glands (salivary glands, liver, gall bladder, and pancreas) in relation to the organs of the digestive tract.

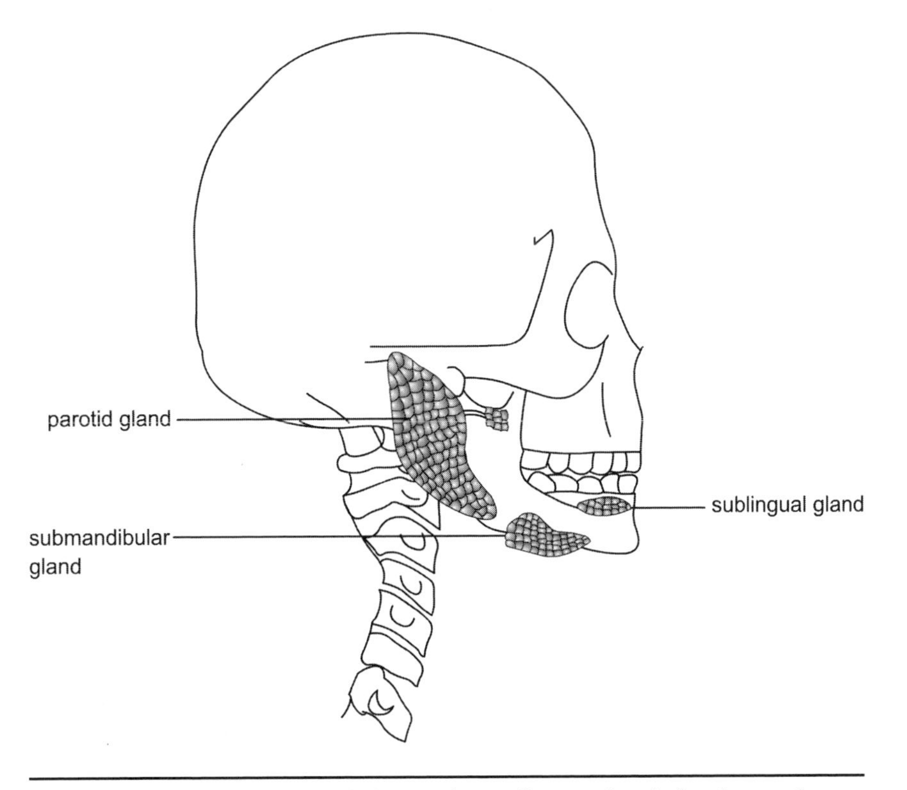

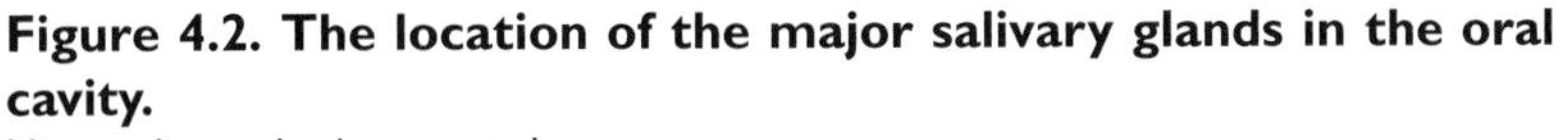
Figure 4.2. The location of the major salivary glands in the oral cavity.
Minor salivary glands are not shown.

speech. The mucus content of saliva serves to protect the tissues of the tongue and cheeks from the action of the teeth. In addition, saliva contains an antimicrobial compound called lysozyme, which serves to reduce, but not completely eliminate, bacterial growth in the mouth. Saliva is slightly alkaline in pH, and thus helps to buffer the oral cavity to the correct pH. The salivary glands also release an enzyme called salivary amylase, which initiates carbohydrate digestion (see Chapters 1 and 2). Each of the salivary glands varies slightly in the content of the saliva, although the saliva from each contains ions, mucus, water, and salivary amylase.

Combined, the salivary glands secrete an average of 1.06 quarts (1,000 milliliters) of saliva daily. The level of saliva production is dependent on a number of factors. In response to dehydration, the body limits saliva production, producing the thirst response. However, it is important to note that the feeling of a dry mouth lags the actual need for water, meaning that a dry mouth signals an advanced stage of dehydration. Most people are aware of increased saliva production in response to the sight or smell of food. This is due to the action of the nervous system, which has the ability to stimulate saliva production based on chemical signals from taste buds or olfactory (smell) glands, or by the touch of food on the tongue. It is also possible to invoke salivation by the memory of food, especially when hungry. The body may also increase or decrease saliva production during illness. In the case of fever or other illnesses, the body may reduce saliva production to conserve water. During times of nausea saliva production may be increased.

GALL BLADDER

The gall bladder is a small sac-like organ, 3.1–3.9 inches (8–10 centimeters) in length, located just under the liver (see Figure 4.3). It is connected to the duodenum of the small intestine by the common bile duct. The gall bladder represents the simplest of the accessory glands in the fact that it primarily serves as a storage location for secretions of the liver (see next section). Unlike the pancreas, liver, and salivary glands, the gall bladder does not produce any chemicals necessary for digestion. Its sole purpose is the storage of bile between meals. Bile is produced by the liver and aids the small intestine in the digestion of hydrophobic molecules such as triglycerides (see Chapter 2).

The common bile duct is actually a conduit from the liver to the duode-

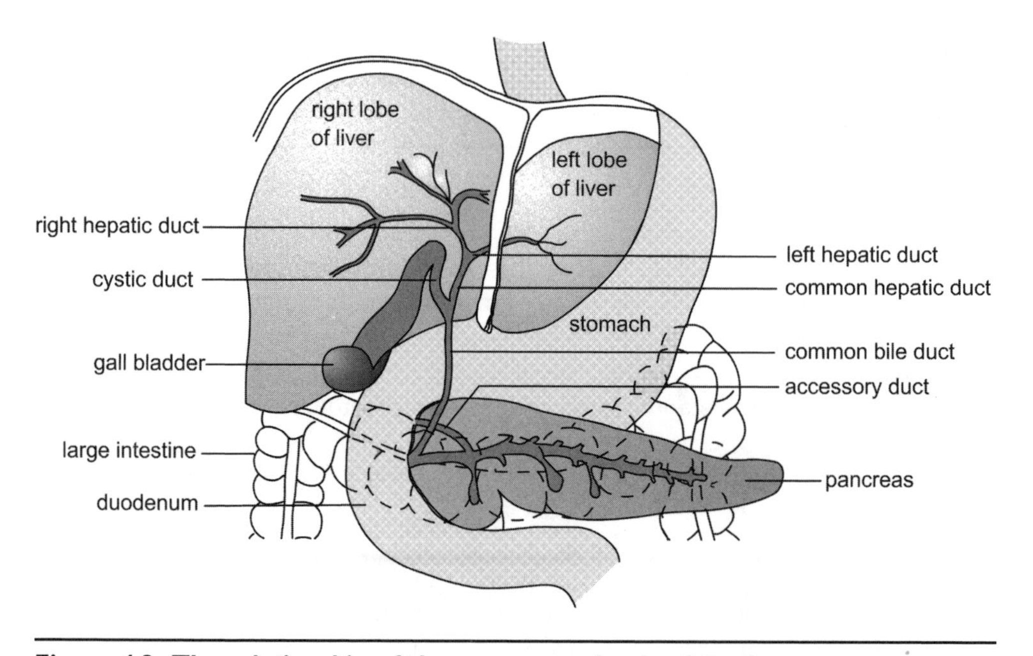

Figure 4.3. The relationship of the accessory glands of the lower gastrointestinal tract to the stomach, small intestine, and large intestine (colon).

num. At the junction of the small intestine is the sphincter of Oddi. When chyme is present in the duodenum, the sphincter of Oddi is open and the bile produced by the liver proceeds directly into the lumen of the small intestine. However, when chyme is absent, the sphincter is closed and the bile being continuously produced by the liver backs up in the bile duct and enters the gall bladder via a small duct called the *cystic duct.* Once stored in the gall bladder, the bile is concentrated and readied for the next meal. When the next meal enters the duodenum, the hormone CCK stimulates the gall bladder to contract, releasing the concentrated bile into the bile duct and into the duodenum.

The gall bladder represents the only nonessential accessory gland. In the case of disease or injury (see Chapter 11), the gall bladder may be removed surgically. When this happens, the bile of the liver proceeds directly into the small intestine and is not stored for later use.

LIVER

The liver represents one of the most important and unique organs of the human body. From a genetic perspective, the cells of the liver are interesting in that some are **polyploid** and **binucleate**. Normally, cells of the body have a single nucleus and are **diploid**, meaning that contain two copies of each chromosome. However, about 50 percent of the hepatocytes in the liver are polyploid cells, meaning that they contain additional copies of each chromosome, while others contain an extra nucleus. There are examples of hepatocytes that have eight or more copies of each chromosome. This arrangement most likely explains the large numbers of organelles found in these cells. Hepatocytes have some of the most abundant endoplasmic reticulum and Golgi bodies of any human cells, which enables them to manufacture large quantities of many biologically important molecules for export. The chromosomal and nuclear state of these cells may also explain regenerative properties of the liver.

With the exception of the skin, the liver is the largest organ in the human body. In an adult, the liver may weigh up to approximately 3 pounds (1.4 kilograms). The liver consists of two primary lobes, called the *right* and *left lobes.* The right lobe, the larger of the two (see Figure 4.3) is sometimes subdivided for study into two additional lobes, called the *quadrate* and *caudate* lobes. The lobes are separated by a **ligament** called the *falciform ligament.* The falciform ligament not only defines the two major lobes, but together with other minor ligaments it helps suspend the liver from the diaphragm. Despite the large size and fairly complex shape of the liver, its tissue is relatively homogenous, which plays an important role in liver physiology.

The liver is also a special organ in that it has the ability to regenerate it-

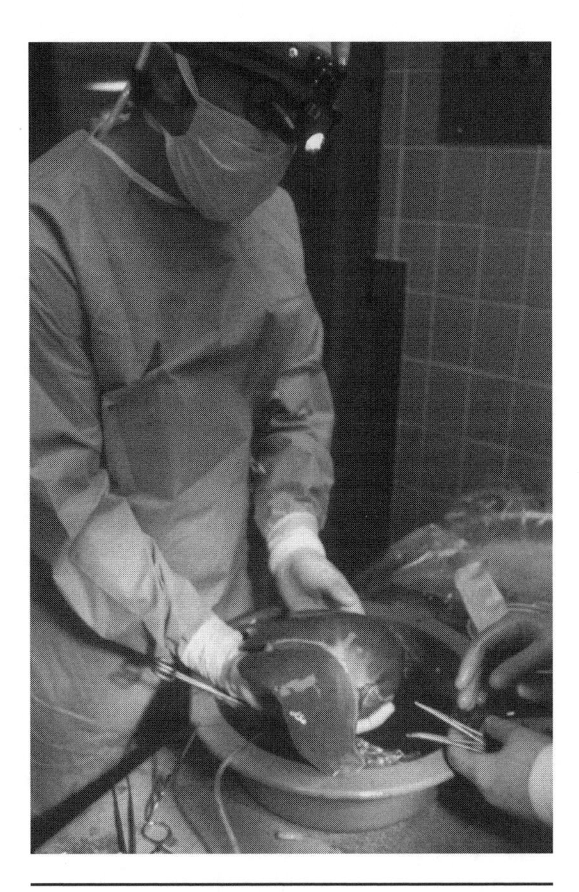

A surgeon holding a healthy liver for transplantation.
© Account Phototake/Phototake—All rights reserved.

self in case of injury or disease. This is primarily due to the redundant structure of liver tissue. Each lobe of the liver consists of self-sufficient subunits called *lobules* (see Figure 4.4). Within each lobule are liver cells called *hepatocytes* and phagocytic cells called *Kupffer's cells*. The purpose of the Kupffer's cells is to engulf worn-out blood cells and invading pathogens, such as bacteria and viruses, arriving from the digestive tract. Thus, these cells technically belong to the immune, and not the digestive, system.

The hepatocytes are the cells where the work of the liver is conducted. Within each lobule, hepatocytes are arranged into hexagon-shaped masses of tissue surrounding a central vein. The liver itself is actually an array of lobules, arranged into the structure of the left and right lobes. This arrangement of tissue means that there are no specialized regions within the liver. Instead, each of these units is capable of performing all of the tasks described below. This redundancy in structure allows the liver to regenerate itself and to sustain substantial damage before failing (see Chapter 11).

On one side of each hepatocyte is a sinusoid. Blood from the digestive system, arriving via the portal circulatory system, enters into the sinusoid cavities. Sinusoids are not the same thing as **capillaries**, but rather represent open spaces from which the hepatocytes can extract nutrients from the digestive system. The phagocytic Kupffer's cells are located along the lining of the sinusoids to protect the liver from pathogens arriving from the digestive tract.

On the other side of the hepatocyte are small vessels called the *bile canaliculi* (bile canals). The hepatocytes continuously produce bile from cholesterol, lecithin (a phospholipid) and bile salts, and secrete it into the bile canaliculi. Within each lobe, these vessels merge into larger structures called the *left* and *right hepatic ducts*, which in turn combine to form the *common hepatic duct*. The common hepatic duct carries bile to the gall bladder, where it then becomes the *common bile duct*. The common bile

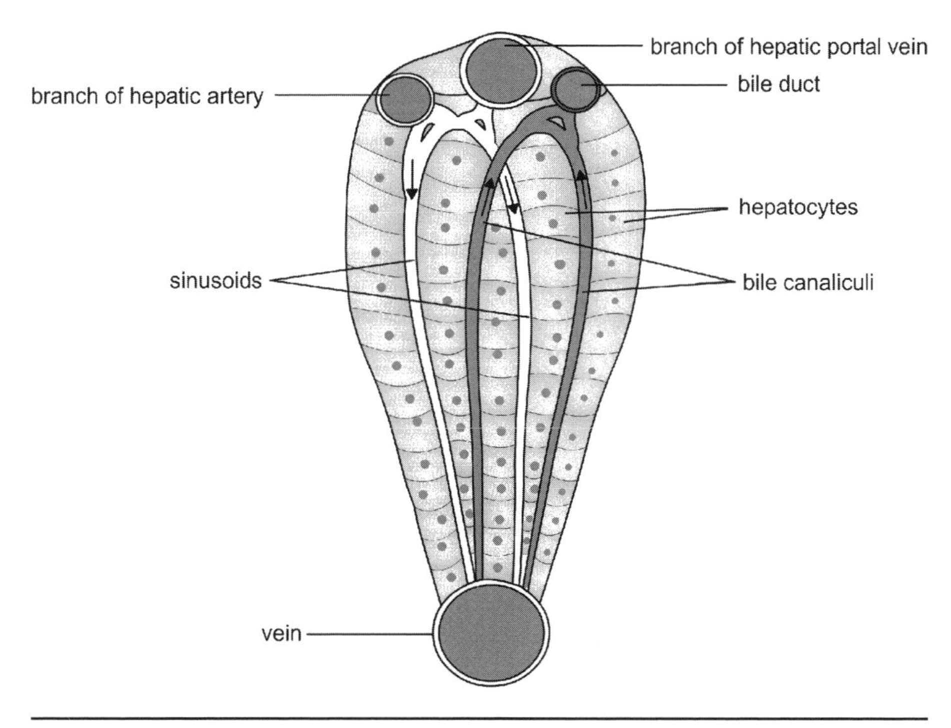

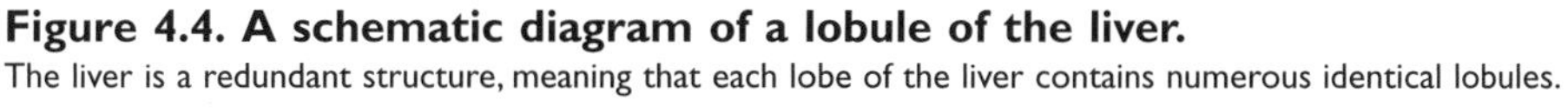

Figure 4.4. A schematic diagram of a lobule of the liver.
The liver is a redundant structure, meaning that each lobe of the liver contains numerous identical lobules.

duct connects with the duodenum of the small intestine through a small valve called the sphincter of Oddi (Chapter 3).

Due to the liver's central role in digestive system physiology, there is a minor deviation in normal blood flow with regard to digestion. Typically, blood leaves the heart via arteries, proceeds to an organ of the body where it enters a capillary bed. The blood then returns to the heart by way of veins. However, in the processing of nutrients there is a minor deviation in this path. Blood leaving the stomach, small intestine, and colon proceeds directly to the liver via the hepatic vein (see Figure 4.5). This minor detour, sometimes called the *portal,* or *hepatic, circulatory system*, ensures that nutrient-rich blood from the digestive tract is first processed and screened by the liver, thus establishing the status of the liver as the master control organ for human digestion. Since the blood from the digestive system is low in oxygen, the oxygen needed for the metabolic functions of the liver cells is delivered by the *hepatic artery.*

The liver plays many important roles in human physiology (see Table 4.2). Since all of the blood leaving the digestive tract first passes through the sinusoids of the liver, the hepatocytes have the ability to screen and filter nutrients and other materials from the blood before it is delivered to the

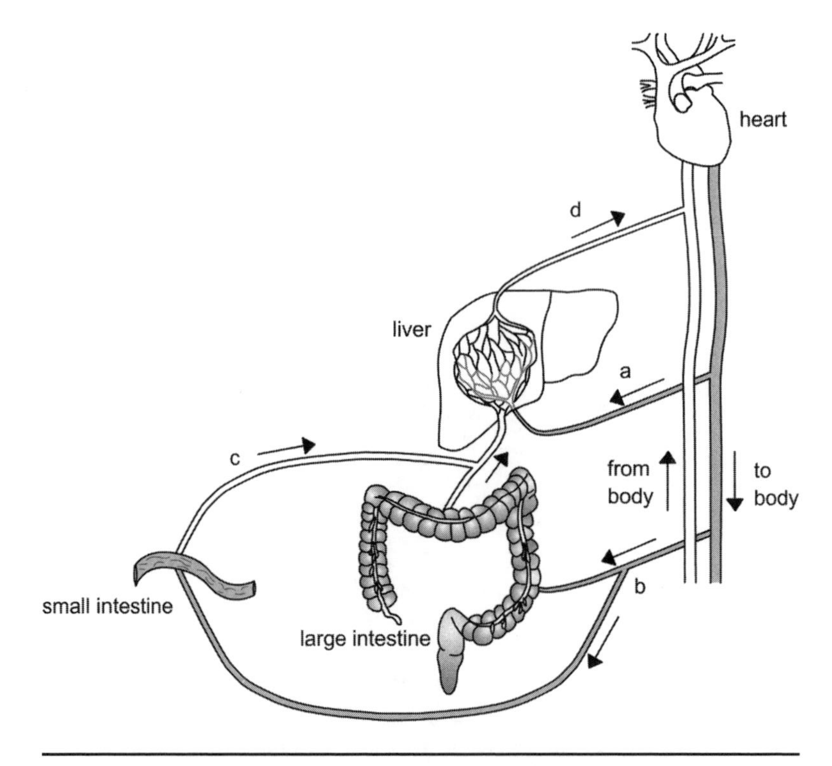

Figure 4.5. Diagram of the hepatic circulatory system.
Blood leaving the heart may proceed directly to the liver (a) or to the intestines (b) by way of arteries. However, nutrient rich blood from the intestines first returns to the liver (c) where nutrients are filtered before the blood returns to the heart (d).

remainder of the body tissues. As a filter, the liver has the ability to remove hormones (see the Endocrine System volume of this series for more information) from the blood. It also can remove drugs or other toxins (such as alcohol) that may be present in the blood. These may either be deactivated chemically, or excreted into the bile. For example, penicillin, a common antibiotic, is removed from the blood by the hepatocytes and excreted into the bile. It eventually leaves the body with the fecal material.

In addition to screening, the liver secretes several important compounds. As previously noted, the hepatocytes manufacture bile salts from cholesterol. This bile is either secreted into the duodenum, or stored in the gall bladder. In addition, the liver excretes a compound called *bilirubin*, which is derived from the destruction of worn-out red blood cells. It is bilirubin that gives bile its characteristic yellow color. In the small intestine, bacteria breakdown bilirubin into *stercobilin*, giving the fecal material its brown color. However, a small amount is reabsorbed by the blood system and eventually excreted by the kidneys. This small amount of bilirubin is responsible for urine's yellow color.

TABLE 4.2. Summary of Liver Functions in Metabolism

General Function	Examples
Detoxification	Removal of steroid hormones from blood Chemical inactivation or excretion of drugs and alcohol
Nutrient Metabolism	Deamination of excess proteins Gluconeogenesis Regulates blood glucose levels Formation of ketone bodies
Synthesis	Manufacture of plasma proteins Manufactures cholesterol and lipoproteins Manufactures bile salts Activation of vitamin D
Miscellaneous	Excretes bilirubin Stores vitamin and minerals Destruction of worn-out white blood cells

The liver also directly interacts with many of the vitamins and minerals needed by the body. The liver acts as part of a three-stage activation system for vitamin D, a hormone-like vitamin manufactured by the skin, kidneys and liver that influences calcium absorption in the small intestine. In addition, the liver may store fat-soluble vitamins such as A, D, E and K, and the water soluble vitamin B_{12}. Within each hepatocyte are proteins called *apoferritin* that actively bind to iron, storing it as a compound called *ferritin*. The liver represents an important storage site for iron. Copper is also stored in the liver, although in quantities much lower than iron.

The liver plays an important role in the metabolic processing of the energy nutrients (lipids, carbohydrates, and proteins). For lipids, the liver is active in the synthesis of lipoproteins, a molecule that transports triglycerides and cholesterol in the circulatory system. The degree that these lipoproteins are loaded with triglycerides determines their density. Lipoproteins are assigned names such as high-density lipoprotein (HDL) and low-density lipoprotein (LDL; see Chapter 1). In addition, the liver plays a role in the manufacture of *ketone bodies* from triglycerides. Ketone bodies are derived by the incomplete **oxidation** of fatty acids, which serve as an alternate energy source for the brain once glycogen stores are depleted in the body, as is the case with starvation. Ketone bodies are acidic and thus alter the pH of the blood, a condition called *ketosis*.

The liver also plays a central role in carbohydrate metabolism. Following a meal, hepatocytes remove glucose from the sinusoids and form the complex carbohydrate glycogen. The liver glycogen reserves are used to maintain blood glucose levels. The liver stores approximately a third of the body's total glycogen stores, with the remainder being held in muscle tissue to power contractions. Muscle glycogen is rarely used to restore blood glucose levels. The regulation of blood glucose levels is under the endocrine control of the pancreas (see following section). During periods of fasting, or when there is inadequate glycogen to maintain blood glucose levels, the liver can undergo a process called **gluconeogenesis**. Gluconeogenesis involves the formation of glucose from non-carbohydrate precursors such as amino acids. In addition, the liver can convert monosaccharides such as fructose and galactose (see Chapter 1) into glucose. Lactic acid, a byproduct of anaerobic energy-releasing pathways of cells (mostly muscle cells) may be converted to glucose by the liver. All of these functions are performed by the hepatocytes.

Proteins are the working molecules of the body (see Chapter 1) and the liver plays an important role in protein metabolism. Proteins are digested in the small intestine to amino acids and are then transported to the liver. To manufacture human proteins, the cells of the body require adequate supplies of all twenty amino acids, yet frequently the food source does not contain the needed levels of each amino acid. Within the hepatocytes, chemical reactions can convert some amino acids to one another, a process called *transamination*. When proteins are needed as an energy source, the liver is responsible for the process of *deamination*, which removes the amino group from the amino acid. Deamination may also be used to rid the body of excess protein. Following deamination, the liver generates urea and releases it into the blood. The kidneys remove and concentrate the majority of the urea for excretion (some is lost through the skin). In addition, the liver generates many of the plasma proteins needed by the blood. *Fibrinogen* and *prothrombrin* are involved in the clotting response of the circulatory system. The *albumins*, which are involved in maintaining the correct water-solute balance in the blood, are produced by liver hepatocytes. In addition, various *globulins* are produced by the liver. Globulins are transport molecules. They are involved in the formation of lipoproteins and the binding of hormones, vitamins, and other nutrients. Adequate globulins are critical for the transport function of the circulatory system.

As indicated, the liver represents one of the most important organs of the human body. Due to its central role in nutrient processing and detoxification, diseases of the liver, such as hepatitis, can be very detrimental, even fatal (see Chapter 11).

PANCREAS

The pancreas is an irregular-shaped gland that is located just below the stomach and adjacent to the duodenum of the small intestine (see Figure

4.3). It averages between 4.7–5.8 inches (12–15 centimeter) in length, and a little over 0.8 inches (2 centimeters) in thickness. For descriptive purposes it is divided into three major sections, although there is little difference in the physiology of the sections. The *head* is located closest to the duodenum and is connected to the digestive tract by two ducts. The *hepatopancreatic duct* is a common duct formed by the linking of the bile duct and pancreatic ducts. A second duct, called the *duct of Santorini*, directly connects the pancreas to the duodenum. Moving away from the duodenum and the head of the pancreas are the regions called the *body* and *tail*.

The pancreas actually represents two separate organs, both of which contribute to digestion, which are integrated into a single structure. A portion of the pancreas is an **exocrine** gland, meaning that it secretes compounds into a cavity. The second major area of the pancreas is the **endocrine** tissue, which secretes chemicals into the bloodstream. In general, the exocrine functions of the pancreas can be described as those directly involved with the processing of nutrients in the duodenum, while the endocrine is best described as those functions that involve hormones and the regulation of glucose **homeostasis** in the body. Both types of tissue exist throughout the pancreas.

Two types of cells make up the endocrine portions of the pancreas (see Figure 4.6). *Duct cells* secrete what is formally called the *aqueous alkaline solution*. This solution is primarily sodium bicarbonate ($NaHCO_3$) and its purpose is to neutralize hydrochloric acid coming through the pyloric sphincter along with the chyme (see Chapter 2). These cells are named due to their close proximity to the pancreatic ducts. Deeper within the pancreas are groups of cells called *acinar cells*. These cells are responsible for generating the enzymatic secretions of the pancreas and together may excrete 1.6–2.12 quarts (1–2 liters) of fluid per day into the duodenum.

The enzymatic secretions produced by the acinar cells of the pancreas contain three basic classes of enzymes. These are the *proteolytic enzymes* (proteins), *pancreatic lipase* (lipids) and *pancreatic amylase* (carbohydrates). It is these enzymes that enable the small intestine to conduct its physiological function as the major organ of digestion and absorption. The pancreatic amylase and lipase enzymes hydrolyze carbohydrates and fats, respectively, into their monomers (see Chapter 1), which are then absorbed by the small intestine (see Chapter 3). However, the action of the proteolytic secretions are a little more complex.

The proteolytic enzymes released by the pancreas are initially inactive. This protects the pancreatic cells from being damaged. The proteolytic excretions of the pancreas contain a mixture of enzymes: *trypsinogen*, *chymotrypsinogen*, *proelastase*, and *procarboxypeptidase*. Once released by the acinar cells, these inactive enzymes proceed through the pancreatic duct into the duodenum. Once in the lumen of the small intestine,

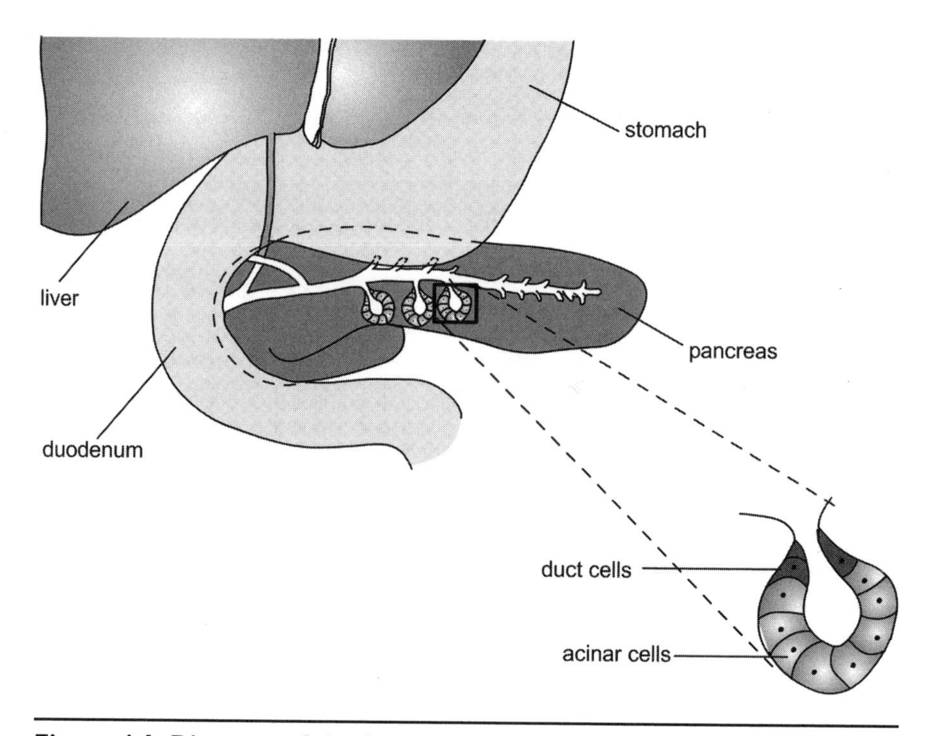

Figure 4.6. Diagram of the internal exocrine structure of the pancreas showing the relationship between acinar and duct cells.

trypsinogen is activated by an enterokinase enzyme located in the mucosal layer of the duodenum. The active form of trypsinogen is *trypsin*. Activated trypsin is an autocatalytic enzyme, meaning that it has the ability to react with additional trypsinogen to produce trypsin. Trypsin also activates chymotrypsinogen, proelastase and procarboxypeptidase to produce *chymotrypsin*, *elastase*, and *carboxypeptidase*, respectively. Together these enzymes are responsible for digesting proteins in the food into individual amino acids or short chain peptides for absorption by the small intestine (see Chapter 3).

The endocrine sections of the pancreas consist of regions called the *islets of Langerhans*. The majority of the cells (approximately 99 percent) in the pancreas are exocrine, but while not as plentiful, the cells within the islets of Langerhans play an important role in glucose metabolism. There are four types of cells in the islets of Langerhans, two of which are involved in glucose regulation. *Alpha-cells* secrete a hormone called *glucagon*, which interacts with the hepatocytes in the liver to increase the hydrolysis of glycogen to glucose, thus increasing blood glucose levels. Glucagon also accelerates the conversion of lactic acid into glucose and the conversion of certain amino acids into glucose by the process of gluconeogenesis. *Beta-*

cells secrete *insulin*, which has the opposite effect. Insulin increases glucose uptake by cells of the body, thus lowering blood glucose levels. Both insulin and glucagons are hormones, and as such are under fine regulation by the body. The endocrine sections of the pancreas are technically part of the endocrine system. Additional information on their function, and diseases of the endocrine areas of the pancreas, may be obtained in the Endocrine System volume of this series.

History of the Digestive System: Ancients to the Renaissance

The history of discovery involving the digestive system encompasses two separate areas of research. First, there are the obvious discoveries of human anatomy involving the digestive organs. These discoveries examined not only the physical structure of the organs, such as their position in the body cavity and their internal connections, but also their relative function to the process of digestion. Second, and just as important, were the investigations into the chemical processing of nutrients by the human body, including the discovery of enzymes and descriptions of enzymatic function. This was the start of the sciences of **organic chemistry** and **biochemistry**, two areas of research that remain active in the modern world. The next three chapters of this text, starting with what is known of ancient science and concluding with recent discoveries, will examine the major contributions of scientists, philosophers, and inventors to our current understanding of the digestive process.

PREHISTORIC AND ANCIENT SCIENCE

While it is unlikely that we will ever know the extent to which cultures before recorded history knew of digestive processes and anatomy, there can be little doubt that ailments of the digestive tract, and local remedies, were known to many cultures across the globe. Eating was an important factor in ancient life (as it is today), and ailments of the gastrointestinal system must have been common before sanitation and food preservation were introduced. However, until the time of the ancient Egyptian civilizations, there are few surviving records of medical procedures or treatments. Most likely, available knowledge in many cultures was passed on verbally from generation to generation.

The ancient Egyptians, whose civilization flourished between 3100 and 300 BCE, are the first culture for which we have records of their study of the digestive system. Although not all records of Egyptian science remained intact until modern times (many were lost due to natural disasters and war), there exists a series of papyrus scrolls that give an indication of the level of the culture. Egyptian science is actually a form of supernatural magic and only remotely would pass for science in today's world. The Egyptians believed that ailments were caused by the gods, and thus all treatments involved a spiritual component. However, their preoccupation with the mummification of the dead, in which the internal organs were removed from the body and stored separately, meant that they were at least in direct contact with the digestive organs. Furthermore, their use of libraries to store important documents meant that information could easily be disseminated among physicians, philosophers, and early scientists, thus greatly increasing the influence of their ideas on latter cultures.

In general, Egyptian medicine in regards to the digestive system focused mostly on the purging the system using mixtures of seeds, oils, beer, and other available chemicals. The Egyptians developed remedies for the treatment of constipation, parasitic worms, hemorrhoids, and indigestion. All of this knowledge was passed on to the Ancient Greeks (see the following section), whose philosophers visited the Egyptian libraries.

Another ancient culture from which we have some records of studies is the Mesopotamian cultures. Starting with the Sumerians and leading to the Assyrian and Babylonian civilizations and centered in the area of the Middle East that is now Iraq and Syria, the Mesopotamian cultures dominated the region starting around 3000 BCE. Although their science, like the Egyptians', focused significantly on mystics and magic, including the practice of **haruspicy**, they did possess some important knowledge of the digestive system. Most of this was recorded not as scientific texts, but as therapeutic tablets that were passed on from one generation to the next. They in particular recognized a number of hepatic ailments, including jaundice (see Chapter 11), and problems associated with constipation, hemorrhoids, and nausea. As with the Egyptians however, the Mesopotamian cultures focused more on treatment than understanding the cause of the ailment.

Studies of digestive system medicine were of course not limited to the West and Middle East. In ancient India, for example, starting several centuries BCE, people believed that the purpose of bile was to cook the food. For this reason they considered bile to be one of the more important fluids in the human body.

Chinese medicine dates back several thousand years BCE. As was the case with other ancient cultures, Chinese medicine focused on the spiritual balance of the body to maintain health. Treatments were typically developed to restore the spiritual balance and not to determine the physiological cause of the problem. The Chinese believed in the major forces that maintained balance and health in humans, the *Yin* and the *Yang*. The Yang represented masculine characteristics, while the Yin was associated with more feminine qualities. The Chinese believed in elaborate interactions between the organs, with organs being related to one another (e.g., mother and father) and having enemy organs within the body. They believed that the lungs, heart, spleen, kidney, and liver were all involved in the process of nourishment. Early in the civilization, surgery was used effectively to treat ailments, but after the development of the religious beliefs of Confucianism around 470 BCE, surgery in general was discouraged since the body needed to remain intact and should not be disturbed. However, **acupuncture** remained a method a treating internal ailments, including those of the digestive system. Some acupuncture needles were up to 10 inches long in order to penetrate the internal organs. Acupuncture is still widely practiced in eastern medicine and is recognized as an alternative form of medicine in Western cultures.

"Acupuncture points on the stomach meridian." Ch'en, Wen—Chih, 1628. © National Library of Medicine.

While each of these major ancient civilizations possessed beliefs and rituals regarding the physiology of the digestive system, few of these had a tremendous bearing on the development of more modern ideas. While it is known that there existed a transfer of knowledge from the east to the west, usually using Arabic cultures as an intermediary, most knowledge was transferred piecemeal, and therefore lacks a comprehensive examination of the digestive system. One exception may be the ancient Egyptian libraries, which were still relatively intact during the early years of ancient Greek society. But in general, most of the ancient theories and ideas passed out of knowledge when the source civilization declined, and thus had to be "reinvented" later.

ANCIENT GREEKS AND ROMANS

The ancient Greeks were well aware of Egyptian science, and it is evident from historical records that many of the prominent Greek philosophers visited the great Egyptian libraries. The true start to the study of the physiology and anatomy of digestion begins with the ancient Greeks (see "Chronology of Some Important Events, 400 BCE to 1600"). For science, the Greeks represent the first group who attempted to explain natural phenomena using scientific explanations. While the worship of multiple deities, or polytheism, was still present, and indeed thrived in Greek culture, Greek science was focused on examining the physical laws of nature. Despite the credit given to the Greeks for the beginning of modern scientific thought, the Greeks were primarily observationists and philosophers. While there are examples of some experimentation being conducted, in most cases the Greeks focused on logic and observation. The dynamic of forming a hypothesis and then testing the hypothesis with planned experiments, the central theme of the modern scientific method, was absent. Therefore, many Greek ideas, while plausible from a logical point of view, actually represent incorrect interpretations of observational data. However, due to the power of Greek culture, and its tremendous influence on the Romans and later European cultures, many of these Greek ideas persisted well into the sixteenth and seventeenth centuries (see Chapter 6). But while their ideas may seem odd in light of modern day studies, their work does represent an important starting point for later studies.

This change in approach to the study of the digestive system is indicated in the early works of Menon (ca. fourth century BCE), a physician who studied under Aristotle. Menon's writings provide one of the first known examples of quantification in the science of medicine. In his work *History of Medicine*, Menon provides details of his own experiment, which attempts to determine the metabolic function of an animal by recording weight gain and loss in relation to food intake.

About a century later, Herophilus (ca. 300 BCE) made some of the first recorded anatomical descriptions of organs of the digestive system. Herophilus did the majority of his studies under the Egyptians at the Museum at Alexandria. Often called the "Father of Anatomy," mostly for his work on the nervous system, Herophilus is responsible for describing and naming the duodenum of the small intestine. The name duodenum comes from the Latin word *duodecum*, meaning "twelve." Herophilus determined that this section of the small intestine was about the length of twelve finger widths, thus the name. Many of Herophilus's anatomy terminology is used in modern medicine. Herophilus is also credited with being the first to describe the lacteals, which although associated with the small intestine, are actually part of the lymphatic system.

Chronology of Important Events, 400 BCE to 1600

ca. 400 BCE	Menon performs the first experimental tests to quantify metabolic rate.
ca. 300 BCE	Herophilus makes one of the first descriptions of the duodenum and lacteals of the small intestine.
ca. 200	Galen develops a model of physiology for the digestive-circulatory system that remains the primary basis of western thought for almost 1,500 years.
1316	Mondino publishes *Anothemia*, indicating a renewed interest in the study of human anatomy in Europe.
1543	Vesalius publishes *The Fabric of the Human Body*, considered by many to be one of the greatest scientific books of all time.

At around the same time and location as Herophilus, Erasistratus (ca. 250 BCE), sometimes called the "Father of Physiology," was developing explanations on how the systems of the body interacted. Once again paying special attention to the nervous and circulatory systems, he did confirm the presence of the lacteals first described by Herophilus. Erasistratus's ideas on human physiology were far ahead of his time and many of which were not "rediscovered" or confirmed for almost two millennia. For example, he discounted the idea that food was cooked in the stomach by describing the physiological function of stomach muscles. He also described how the pharynx served as the junction between the respiratory and digestive systems, and how it closed off the respiratory system while swallowing. In is interesting to note the progress that both Herophilus and Erasistratus were making on the anatomy and physiology of the human body. Unfortunately, by the end of the third century BCE there was a change in Egyptian philosophy. Egyptians came to believe that the body must be kept intact to reach the afterlife, effectively ending the study of human anatomy for almost two thousand years.

This inability to study humans directly meant that those interested in human anatomy and physiology were forced to study other vertebrates. While in many cases there are similarities in the anatomy and physiology of vertebrates, there also exists a tremendous amount of variation, especially in the structure of the digestive system. For example, the physician Rufus of Ephesus (ca. 50 CE) examined the structure of a canine liver and concluded that it possessed five distinct lobes. Since the structure of the human liver could not be determined directly, it was widely accepted for almost 1,700 years that the human liver possessed five lobes instead of the actual three.

By about 250 BCE, Greece came under the control of the Roman Empire. Although the Romans are not directly recognized for their contributions to science, pockets of Greek researchers continued to work on scientific problems under the governance of the Romans. Perhaps the greatest philosopher of medicine during this time was the Greek physician Galen (129–216 CE), who for a time served as the physician to the Roman emperor Marcus Aurelius. Galen's numerous works have almost no basis in experimental facts, although his writings suggest that he supported experimental processes. Rather he focused primarily on descriptive and philosophical studies of the human body. Thus, while his writings and position in the medical community earned him the reputation as a scholar of science, many of his ideas were scientifically flawed. However, in most cases they remained the prime source of medical information until well into the sixteenth century.

To Galen, the digestive and respiratory systems were invariably linked to one another. Without the ability to examine human anatomy firsthand, Galen attempted to link together the ideas of Aristotle (384–322 BCE) and Erasistratus (304–250 BCE) into one coherent model of how the body functioned. Erasistratus had observed a milky fluid, called *chyle*, in the lacteals of the small intestine and concluded that this was the food leaving the gastrointestinal tract. In actuality, this is only the lipid-based nutrients, the soluble substances heading to the liver via the hepatic vein (see Chapter 4). However, being part of the lymphatic system, the lacteals do not directly connect to the liver, and thus Galen lacked a means of getting the nutrients to the liver. To remedy this, he also moved the chyle to the liver via the portal vein, which actually carries blood. Aristotle believed that the liver was the site where the blood was manufactured and it was this blood that was consumed by the tissues of the body for growth and development. In Galen's model, once the chyle was received in the liver, it was used to manufacture the blood. The role of the liver in blood manufacture was widely supported by the observation that the liver is frequently the bloodiest organ of the human body. In Galen's model, the liver became the source of the "vegetative spirits," proposed by both Aristotle and Plato (427–347 BCE), which were responsible for growth. To finish the trilogy of "spirits," the heart provided the "vital spirits," or soul, and the brain provided the "animal spirits," or controlling force. The fact that the liver filtered nutrients from the blood, not manufactured it, would take centuries to determine.

One can question why these ideas persisted for such a lengthy period of time, when simple observations of human biology would indicate that the chyle does not proceed directly to the liver, the portal vein carries blood, and the liver does not manufacture blood. A number of theories have been proposed, but the basic cause was probably the inability to examine the human body directly. A number of historians have pointed out that in most European and Arabic medical institutions, students were not encouraged to directly observe human anatomy until the eighteenth and nineteenth centuries. Instead, lecturers described processes and recited Galenic ideas and principles. This stagnated medical thought and inhibited advances in understanding basic physiology.

DARK AGES

Galen's death effectively marked the end of the study of human anatomy and physiology in the west for over one thousand years. Not only did this time lack any significant advances in the study of human biology, or almost any other science for that matter, but there developed a widespread mistrust of scientific information so that much of what had previously been achieved was destroyed. However, some information remained, housed mostly in the Arabic world, which served as a form of sanctuary for scientific information until Europe began to emerge from the Dark Ages.

While Europe stagnated, Arabic medicine was flourishing. Galen's works were widely translated into Arabic, and in many ways served as a foundation for much of the early work in medicine in the Arab world. As was the case with mathematics and chemistry during this time, the Arabs added a significant amount of new ideas to the Greek foundation. Arabic physicians, such as Abu Bakr al-Razi (ca. 854–925), also known as Rhazes; ibn Sina (Avicenna, 980–1037), the "Galen of Islam"; ibn Rushd (Averroës, 1126–1198); and ibn al-Quff, all made substantial contributions to the understanding of human physiology and medicine. Some of these, such as the discovery of capillaries, occurred centuries before the official "discovery" occurred in western Europe.

By the thirteenth and fourteenth centuries Europe was emerging from the Dark Ages, and with this transformation began a slow development of interest in scientific research. At the start of this period dissections of humans are still relatively rare. However, some medical schools, most notably the Surgical School at Bologna, were once again initiating the practice. The start of modern descriptions in anatomy is often to said to have begun with the work of Mondino de' Luzzi (ca. 1270–1326) in 1316. In that year Mondino, as he is frequently called, published the book *Anothemia*, which contained some detailed descriptions of the human body. Mondino was greatly influenced by the Arabic anatomists, and in many cases he did not expand significantly on their work. He divided the organs of the body according to the Greek concept of "spirits." Thus the *natural members*, generators of the natural (or vegetative) spirits, consist of the digestive organs, such as the liver and stomach. Although his descriptions were somewhat detailed, he continued to make some of the same mistakes as earlier scholars. He described the liver as having five lobes (versus the actual three), which brings into question if he was actually examining human, and not canine, specimens. Mondino was one of the first to describe the pancreatic ducts leading to the small intestine and performed one of the first detailed descriptions of the gall bladder. While obviously not completely accurate, Mondino's work does represent an important first step in developing descriptions of human anatomy.

RENAISSANCE

From a historical perspective, the Renaissance is said to have begun in the fourteenth century in Italy. Mondino's work therefore represents the transition between the suppressed science of the Dark Ages and the rebirth of science and art in the Renaissance. The Renaissance marks not only a renewed interest in scientific research, but also the combination of art and science into a single field of study. Perhaps the most famous, and possibly influential, example of the combination of science and art was performed by the Italian inventor and artist Leonardo da Vinci (1452–1519). While not truly a scientist, Leonardo da Vinci was more of a scientific illustrator, as were many artists of this time. (Michelangelo [1475–1564] and Raphael [1483–1521] also recorded aspects of human anatomy in their paintings and drawings.) Leonardo da Vinci is known to have dissected over thirty cadavers, producing detailed drawings of all aspects of human anatomy. In many cases, the quality of these drawings would not be surpassed for centuries. Unfortunately, Leonardo da Vinci did not publish his ideas, and thus the quality of his work, which would have greatly assisted the development of anatomy in the Renaissance, remained undiscovered for centuries. Still, it does demonstrate the level of interest that was developing during this time in Europe.

The ideas and models of human physiology presented by Galen in the second century CE remain the standard for study until the time of the Renaissance. The individual who is most often credited with ending Galen's reign in science is Andreas Vesalius (1514–1564). Vesalius was a gifted scientist who became a physician at the young age of 23. Initially, Vesalius followed the training and beliefs of Galen. One of Vesalius's earliest books described a five-lobed liver and Galen's hepatic circulatory system. However, Vesalius was immensely fond of research, and whenever possible he was active in dissecting both human cadavers and animals. Local magistrates made the bodies of executed criminals available to Vesalius, which greatly enhanced his ability to conduct research. He also frequently lectured, and often conducted dissections live in the classroom while lecturing.

Over time, Vesalius began to find fault with many of Galen's ideas. Once again these were primarily focused on the circulatory system, specifically the structure of the human heart. However, as his doubt grew, he was encouraged to record and publish the results of his research. In 1543 he published *The Fabric of the Human Body*, a 600-page series of books that included seventeen full-page drawings of human anatomy. The fifth book contained a detailed examination of the digestive, and other abdominal, organs. This book contained descriptions of the gall bladder and layout of the intestines; while it illustrated the presence of an appendix, there was no attempt to explain its function.

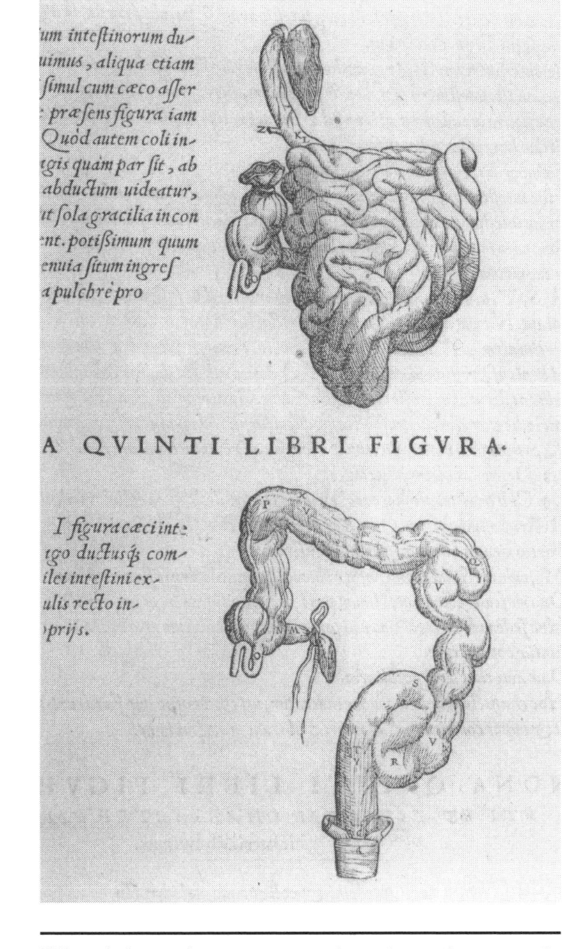

"The abdominal cavity exposed to show the stomach, liver and intestines." © National Library of Medicine.

The content of *The Fabric of the Human Body* was so exact and well-researched that it influenced an entire generation of anatomists. Many historians consider *The Fabric of the Human Body* to be one of the greatest scientific publications of all time. The acceptance of Vesalius's ideas ended the majority of the existing support in Galen's ideas and effectively started a new era of study in anatomy. Other anatomists, such as Realdus Colombus (ca. 1510–1559), Gabriele Fallopio (ca. 1523–1562) and Bartolommeo Eustachio (1520–1574), followed Vesalius's lead and greatly advanced the study of other body systems.

The science of anatomy experienced some tremendous highs and lows in the 2,000 years between the first recording of ancient science and the start of the seventeenth century. If science had progressed at the pace originated by the Egyptians and Greeks, then major discoveries and advances definitely occurred earlier in history. Yet the pace of science in these times was directly tied to the cultural climate. The inability to examine the human body directly, the skepticism of scientific processes, and a lack of research facilities greatly inhibited detailed examination of human biology, including the digestive system.

The work of Vesalius and others marks an end to this era and in the seventeenth century, science and religion begin to become separate entities, thus marking the beginning of the Scientific Revolution and more comprehensive studies of human anatomy and physiology.

History of the Digestive System: Seventeenth to Nineteenth Centuries

The time between the seventeenth and nineteenth centuries marks a defining time in the study of human anatomy and physiology. In the early seventeenth century, the work of scientists such as William Harvey (1578–1657), Galileo Galilei (1564–1642) and Johannes Kepler (1571–1630) shook the foundations of ritualistic beliefs that had existed since the time of the ancient Greeks. Early seventeenth century scientists began to recognize that the scientific method, with its focus on experimentation and logic, was the basis of sound scientific thought. Although research in the seventeenth century focused primarily on the physical sciences of astronomy and physics, there were some important advances in physiology during this time. By the end of the seventeenth century, a rift was forming between the chemists and the alchemists, which eventually gave rise to the study of empirical chemistry in the eighteenth century. This development was crucial for subsequent discoveries of chemical reactions within the body. The nineteenth century represents the culmination of two centuries of work. During this time, scientists not only had the interest in exploring the human body, but also the tools to do so effectively.

The three centuries covered in this chapter set the stage for scientific advances of the twentieth and twenty-first centuries. The twentieth century represents the century of greatest growth in medicine and human biology. In order to understand modern medicine and anatomy and physiology, it is first necessary to examine the contributions of the pioneers in the seven-

Chronology of Important Events, Seventeenth to Nineteenth Centuries

1614	Sanctorius begins his experiments on metabolism using the "weighing chair."
1628	Harvey's study of blood circulation removes the liver as an organ of circulation.
1662	Steno identifies the duct connecting the salivary glands to the oral cavity.
1664	de Graaf isolates pancreatic juice.
1736	von Haller identifies the role of bile in fat digestion.
1752	Réaumur's experiments with digestion indicate that chemical activity is involved.
1820	Branconnot discovers glycine, the first amino acid to be directly associated with a protein.
1824	Prout identifies the acid in gastric juice to be hydrochloric acid.
1825	Chevreul discovers the structure of triglycerides.
1833	Beaumont studies the physiology of digestion *in vivo* using a human patient.
1836	Schwann identifies pepsin.
1855	Bernard discovers the glycogenic abilities of the liver.
1857	Bernard discovers glycogen.
1869	Miescher discovers the nucleins, the modern day nucleic acids.
1886	Kiliani discovers the structure of glucose.
1888	Pavlov develops the theory of conditioned reflex as a result of his studies of the digestive system.

teenth to nineteenth centuries (see "Chronology of Important Events, Seventeenth to Nineteenth Centuries").

SEVENTEENTH CENTURY

During the seventeenth century, there was not a coordinated effort to examine the action of the digestive system as a whole. Instead, investigators were piecing together aspects of human physiology from a number of different perspectives. A significant amount of research was continued by scientists studying the comparative anatomy of other vertebrates. From the time of the Renaissance, anatomists were using animals for anatomy studies, due to their inability of obtaining human cadavers for study. While works based on such studies represented an important advance in the science of **zoology**, they did little to further the understanding of the human digestive system. However, they did generate a tremendous desire on the part of scientists to expand their investigations into humans. As the seventeenth century progressed, the academic climate became more favorable, and some studies were conducted on human cadavers, although these tended to focus on the circulatory system at first. Still, the information that was collected undermined the prevailing thoughts that had persisted since the time of the ancient Greeks.

For example, since the time of Galen (see Chapter 5), the liver was believed to be the organ that was responsible for the production of blood and the mixing of the blood with "vital spirits," an early form of nutrition. The blood was then mixed with more "vital spirits" in the heart and consumed by the tissue. The actual role of liver in digestive processes (see Chapter 4) was not immediately recognized. With the discovery of blood circulation by William Harvey in 1628, the accepted role of the liver as a circulatory organ diminished, and it began to take a prominent role in studies of digestion and nutrition.

Despite the outdated nature of Galen's theories on the human body, many of his ideas persisted until the seventeenth century. Galen believed that the role of the stomach was to cook the blood to an elevated temperature prior to it being delivered to the tissues. A more physiologically correct role for the stomach was proposed by Jan Baptista van Helmont (1580–1644), a Belgian chemist. Van Helmont belonged to a developing group of chemists called the *iatrochemists.* Iatrochemists attempted to apply chemical principles to the study of human physiology. They were in fact the forerunners of modern-day biochemists. Van Helmont believed that the stomach was the site of a fermentation process that released nutrients from food. He also correctly concluded that the gall bladder was involved with digestion, although he incorrectly assigned it the function of neutralizing stomach acid, a process that is actually conducted by the secretions of the pancreas (see Chapter 3). Van Helmont was also incorrect in his idea that the digestive process was controlled by an organ called the *archeus.* (This organ was believed to control all physiological activity in the body.) Unfortunately for Van Helmont, no such organ exists. Instead, digestive control is regulated by the liver, with the hypothalamus and thyroid providing metabolic regulation.

One of the first studies of human metabolism during the Scientific Revolution was conducted by the Italian scientist Sanctorius (1561–1636). Sanctorius designed a special "weighing chair" in an elaborate attempt to link food intake with metabolic processes. For almost three decades, he measured the weight of all food and liquids he consumed, as well as all fecal material and urine he generated. While there were definite problems with his exper-

"Sanctorius in his Balance." Engraving by J. Bengo. © National Library of Medicine.

imental approach, most notably his inability to measure the heat or carbon dioxide gas generated by metabolic functions, he did conclude that the differences in the intake and output measurements were due to a process called "insensible perspiration."

Another iatrochemist, the Dutch physician Franciscus Sylvius (1614–1672), was one of the first to identify the role of the salivary glands, pancreas, and gallbladder as accessory glands to the digestive process. He built on the work of anatomists who were identifying structures of the digestive system. For example, the Danish scientist Nicolaus Steno's (1638–1686) discovery of the duct linking the parotid salivary gland to the oral cavity in 1662 indicated that these structures may be involved in the digestive system. The Dutch scientist Regneir de Graaf's (1641–1673) isolation of pancreatic juice from a dog in 1664 made a similar connection for the pancreas. However, neither determined the purpose of these secretions. Sylvius believed, as did many iatrochemists, that human diseases were the result of chemical imbalances in the body and recognized the role of these glands in maintaining chemical balance. While Sylvius was on the right track, the chemical basis of accessory gland function would not be determined until the eighteenth and nineteenth centuries.

The seventeenth century was not limited solely to the work of the iatrochemists. From the more philosophical perspective, there was a growing belief in this century that investigations of human physiology and anatomy should be focused more on mechanical processes, recognizing that the human body is the sum of its individual parts. A leading proponent of this mechanist view of human physiology was the French philosopher and mathematician René Descartes (1596–1650). Descartes defined many responses in the body as being either voluntary or automatic, which is a correct interpretation of the modern view of how the nervous system is divided.

He correctly assumed that functions of the digestive system, specifically the peristaltic contractions of the intestines, were automated responses. It is now recognized that this is due to the action of smooth muscle lining the gastrointestinal tract (see Chapter 2).

By the end of the seventeenth century the revolution in chemical thinking was well underway, and these studies of chemistry were developing the tools to examine the physiology of the digestive system in greater detail.

EIGHTEENTH CENTURY

While the seventeenth century was the time of great advances in the study of physical sciences such as astronomy and physics, the eighteenth century marked the development of the sciences of chemistry and zoology, or the study of animals. While the majority of the work by zoologists during this time was focused on the classification of animals, the work of the comparative anatomists continued. As was the case with the research of the seventeenth century, there were some important advances in the study of the digestive system, although there still lacked a coordinated effort to examine the system as a whole.

One of the more important advances in digestive system physiology during this time was conducted by the French physicist Ferchault de Réaumur (1683–1757) around 1752. Réaumur belong to a group of physicists called the iatrophysicists. Iatrophysicists believed that all physiological processes in the body could be explained by physical laws, much like the iatrochemists and chemistry. With regards to digestion, Réaumur believed that there were three possible explanations for digestion. Material could be digested by mechanical interactions between the food and the wall of the stomach; by a natural decaying or putrefaction of the food; or by a chemical interaction between the food and the secretions of the stomach. While these ideas all existed before Réaumur's work, what makes his research important is that he designed an experimental system to test which hypothesis was correct.

For his experimental system, Réaumur chose a living hawk. When hawks eat they frequently regurgitate undigested material. For his experiments, Réaumur placed small pieces of meat and vegetable material inside metal containers. At the ends of the containers were small metal grates that allowed fluids to pass into the chambers. He then coaxed the hawk to swallow the tubes. After a few minutes, the hawk naturally regurgitated the tubes. Inside the tube, Réaumur discovered that the meat was partially digested, while the vegetable material was unharmed. Since the meat was not able to come into contact with the stomach lining, mechanical digestion was ruled out; since the food was only in the stomach for a short period of time, it was unlikely that natural decay was the digestive force. That left chemical digestion as the only remaining hypothesis.

Inside the canisters with the meat was an acidic fluid. This fluid did not appear to digest the vegetable material, only the meat. It is now known that gastric juice contains the enzyme pepsin (see Chapter 2) and that the stomach is the initial site of protein digestion. However, Réaumur wanted to know more about this fluid, so he designed a system by which small sponges, tied to fine strings, were introduced into the stomach of the hawk and then retrieved. By wringing out the sponges, Réaumur was able to obtain enough of the gastric juice for study. In another experiment he placed meat into two dishes. One dish contained stomach juice (gastric juice), while the other was left open to the air. The meat in the gastric juice showed signs of digestion, but not decay, while the meat left exposed to the outside was beginning to putrefy. Thus it was the gastric juice that was responsible for digestion in the stomach. Additional experiments with sheep and dogs yielded similar results.

Réaumur's work encouraged other scientists to take a more detailed look at stomach secretions. One of these was the Italian biologist Lazzaro Spallanzani (1729–1799) who greatly expanded on Réaumur's work. Using the same basic method to remove gastric juices from organisms as did Réaumur, Spallanzani exposed tubes containing meat and gastric juice to human body heat by carrying the tubes under his armpits for several days. He obtained the same results as Réaumur, further demonstrating that the action of digestion inhibited natural decay and that the two were separate forms of reaction.

There were other, although relatively minor, discoveries in the eighteenth century. For example, Albrecht von Haller (1708–1777), a Swiss physiologist who primarily researched the nervous system, was one of the first to correctly determine the association of bile with fat digestion. These discoveries, great and small, provided a solid foundation for the substantial research on the digestive system in the nineteenth century.

NINETEENTH CENTURY

In the nineteenth century, the advances in chemistry from the previous two centuries began to have their influence on the study of digestive processes. In Chapter 1, we examined the basic biomolecules that the digestive system must process in order to supply the tissues of the body with the necessary nutrients for growth, reproduction, and physiological function. In the nineteenth century the studies of organic chemistry were allowing scientists to examine the structure of carbohydrates, fats, proteins, and nucleic acids. As noted earlier, this began the study of biochemistry.

The processing of proteins is an important task of the digestive tract since proteins are the working molecules of the cell, with almost all metabolic processes involving at least some level of protein action. In the early nineteenth century scientists had begun to isolate a number of compounds from living tissues that when heated did not melt, but rather assumed a solid form. This class of molecules was initially given the name *albuminous*. Included in this class were proteins such as albumin from egg whites, globulins from blood and a milk protein called *casein*. However, their role in living tissues was not understood. At around the same time, chemists were isolating another group of compounds, which are now called the amino acids. Two of these, asparagine and cystine, were discovered in 1806 and 1810, respectively, as scientists were studying the chemical composition of living cells. While amino acids are now known to be the building blocks of proteins (see Chapter 1), in the early nineteenth century the connection between amino acids and proteins had not been established. The first amino acid that was clearly identified as being associated with the albuminoids was glycine. In 1820, Henri Branconnot (1781–1855) isolated a sweet compound from animal tissue that he thought to be a form of carbohydrates. Upon further investigation he discovered that the compound contained nitrogen, an element that is absent in sugars. At the time the only molecules that were known to contain nitrogen were the albuminoids. The discovery of other amino acids, such as leucine and tyrosoine, followed. While these newly discovered amino acids obviously had some function in living cells, their role in the formation of a functional protein was unclear.

The relationship between amino acids and proteins began to take shape around 1839 with the work of the Dutch chemist Johannes Mulder (1802–1880). Mulder was investigating the albuminoids when he discovered what he thought to be a common empirical formula, $C_{40}H_{62}O_{12}N_{10}$. Mulder believed this to be the structural backbone of all albuminoids, to which other elements such as sulfur and phosphorous could be attached. Since these molecules had the ability to deliver all of the major elements (carbon, hydrogen, nitrogen, oxygen, sulfur, and phosphorous) needed for living cells, Mulder and the Swedish chemist Jöns Berzelius (1779–1848) named the compounds *proteins*, from the Greek word *protos*, Meaning "first importance." By this time, other chemists had begun to work out the structural formula for the amino acids. In their investigations, it was noted that all amino acids consist of the same core structure, with variations attached to the central, or alpha, carbon. Furthermore, advances in chemical procedures toward the end of the century had made it possible to isolate amino acids directly from proteins, so what was missing was the mechanism by which proteins were formed. This was determined in 1901 by the German chemist Emil Fischer (1852–1919). Fischer demonstrated that amino acids are formed by a condensation reaction between the amino end of one amino acid and the carboxylic acid end of a second amino acid (see Figure 1.4).

Proteins were not the only biomolecules studied in the nineteenth century. The class of biomolecules called the carbohydrates was recognized well before the nineteenth century as being an important energy source for living organisms. In the nineteenth century, Fischer was studying the structure of glucose, a simple sugar with the empirical formula $C_6H_{12}O_6$. The German chemist Heinrich Kiliani (1855–1945) had previously determined the structure of glucose in 1886. However, there were other sugars known to have the same chemical formula as glucose, such as mannose and fructose. Fischer developed chemical methods to study the structure of these compounds. He discovered that, although these molecules have similar empirical formulas, they possess different structures. These types of compounds are called **isomers,** and their discovery was a significant breakthrough for biochemistry. For his work, Fischer received the 1902 Nobel Prize in Chemistry.

Another of the energy molecules, the fats, also received attention from the scientific community in the nineteenth century, although from a slightly different perspective than the other biomolecules. Like the other biomolecules, scientists in the early part of the century had been trying to determine the structure of fats and lipids. Around 1825, the French chemist Michel Chevreul (1786–1889) determined that the fats actually consisted of two components: glycerol and fatty acids. In his studies Chevreul isolated several different forms of fatty acids, including the common oleic, palmitic and stearic fatty acids.

Undated photo of Emil Fischer in his laboratory. © National Library of Medicine.

In 1869, the Swiss chemist Johann Miescher (1844–1895) detected a unique group of chemicals in the nucleus of the cell. Unlike the other classes of biomolecules, these chemicals contained both phosphorous and nitrogen. These same compounds were also detected in the nuclei of other cells. This class of compounds was initially called *nuclein.* In 1885 the German biochemist Albrecht Kossel (1853–1927) determined that there were actually two different types of nuclein, one which was made of protein and a second which was not a protein.

The chemicals that make up this second type are called the nucleic acids. Kossel received the Nobel Prize in Physiology or Medicine in 1910 based partly on this work. However, little additional work was done on nucleic acids in the nineteenth century, and their importance as the nutrient class responsible for the genetic material was not determined until the next century. Still, by the end of the nineteenth century chemists had isolated all of the major nutrient classes needed for human health.

The work of eighteenth-century scientists such as Réaumur and Spallanzani had proven that digestion and putrefaction were different processes. What was lacking was a means by which digestion occurred. The acidic component of gastric juice—hydrochloric acid—was first isolated in 1824 during studies of digestion by William Prout (1785–1850). But hydrochloric acid itself can't break down proteins completely. In 1837, Jöns Berzelius, a Swedish chemist, described a type of compound that accelerated chemical reactions, without themselves being used up in the process. He called these *catalysts*. Many chemical reactions utilize catalysts, but what interested investigators of digestive processes were catalysts that worked on organic molecules. The study of organic chemistry and fermentation pathways was a major area of research in the mid nineteenth century, and what interested these scientists most were the type of catalysts called the *soluble ferments*.

The first soluble ferment was isolated by Anselme Payen (1795–1871) and Jean Persoz (1805–1868) in 1833, who discovered a catalyst from barley called *diastase*. This enzyme is involved in carbohydrate digestion. Other carbohydrate catalysts, such as **zymase** and **invertase** were also discovered about this same time. For animals, the first enzyme to be discovered was pepsin, the protease enzyme of the stomach (see Chapter 2). This was first discovered by the German scientist Theodor Schwann (1810–1882) in 1836 when he rinsed the interior lining of a stomach and used this extract to digest protein. Pepsin is the catalytic agent responsible for the eighteenth century experimental observations of Spallanzani and Réaumur. The term enzyme to describe these molecules was first used by Wilhelm Kuhne (1837–1900) around 1878.

In addition to these studies of molecules, there was some important research being conducted in the area of digestive physiology. One of the more interesting—and revealing—studies in the history of medicine was conducted between 1822 and 1833 by the American physician William Beaumont (1785–1853). One of Beaumont's patients, a man by the name of Alexis St. Martin, had accidentally been shot in the abdomen. Although Beaumont treated the injury, it did not heal correctly and St. Martin developed a small hole, or fistula, between the outside environment and the internal lumen of his stomach. Beaumont capitalized on this abnormality by using St. Martin as a testing ground for physiological studies of digestion. By tying various foods to a piece of string, Beaumont was able to examine how the digestive juices of the stomach reacted. He not only varied the types of foods, but also the time that food remained in the stomach. He was one of the first to recognize that secretions of gastric juices increase when food is present.

One of the more important scientists to contribute to understanding of the digestive system in the nineteenth century was the French researcher Claude Bernard (1813–1878). For his time, Bernard was a sensational physiologist. His contributions toward the understanding of the circulatory system, respiratory system, and the effects of poisons on the body, or toxicology, are numerous. Starting around 1855, Bernard began a series of experiments on how the vertebrate body handles sugar. Using dogs as the model system, Bernard fed dogs first a diet rich in sugar, and then a sugarless diet. In the first case, sugar was present in both the portal vein (GI tract to liver) and hepatic vein (liver to heart). This demonstrated that the sugar from the diet passed through the liver and into the circulatory system of the body. In the second case, when the dog was given a sugarless diet, sugar was of course missing from the portal vein, but was present in the hepatic vein. This indicated that the liver possessed the ability to manufacture sugar when it was absent in the diet. Bernard had discovered the glucose regulating properties of the liver. In 1857 he isolated glycogen, the animal storage carbohydrate, from the liver. His research was the first step in the study of gluconeogenesis, or the manufacture of new glucose from non-glucose precursors (now known to be protein). Bernard also expanded on de Graaf's earlier work by demonstrating that the pancreas secretes chemicals into the duodenum when food is present in the gastrointestinal tract, but not when it is empty. He learned that one function of these secretions is to break down maltose (a disaccharide) into glucose. His work set the stage for numerous twentieth century studies. In 1869, Paul Langerhans (1847–1888) discovered microscopic structures with the pancreas, now called the Islets of Langerhans. These structures, while not involved in the production of pancreatic juice (see Chapter 4), are crucial to the endocrine control of blood-glucose levels.

In 1888, the Russian scientist Ivan Pavlov (1849–1936) proposed that digestion was a three-stage process consisting of a nervous stage, pyloric stage, and finally an intestinal stage. The nervous stage was a response to a stimulus, or what is sometimes called a conditioned reflex. He was one of the first to propose that the gastrointestinal tract was actually controlled by the nervous system. To test his hypotheses, he trained a dog so that every time a bell was rung the animal was fed. Over time, the dog increased its salivation in response to the ringing of the bell, even if no food was present. The concept of Pavlov's conditioned reflex is still used in psychiatric studies, but it has its routes in the biological sciences. The experimental proof of the biological nature of this response was developed in the early twentieth century (see Chapter 7).

The period between the start of the seventeenth century and the end of the nineteenth century marks a time of great change for the study of the digestive system. Initially working to dispel the incorrect teachings of the ancient Greeks, over the next two centuries physiologists not only developed an understanding of the relationship between digestion and chemistry, but also established some of interconnections between the digestive and endocrine and nervous systems. This research would prove to be crucial for twentieth century studies that not only expanded on the anatomy and physiology of the system, but began to take a close look at diseases of the digestive tract.

History of the Digestive System: Twentieth Century to the Present

During the twentieth century, the sciences of medicine and physiology came of age. Probably at no other time in history have the life sciences experienced more growth and achievements than in the past one hundred years. Advances in three primary areas of study played an important role in advancing the study of human biology.

In the year 1900, three investigators—Erich von Tschermak (1871–1962), Carl Correns (1864–1933), and Hugo de Vries (1848–1935)—independently rediscovered the mathematical principles of inheritance established by Gregor Mendel (1822–1884) in the mid-1800s. While Mendel's work had very little influence on the science of the nineteenth century, the scientific climate had changed drastically by the start of the twentieth century. Charles Darwin's (1809–1882) theory of natural selection had sparked a tremendous interest in the natural sciences, specifically evolution, and many researchers were beginning to wonder how inheritance at the cellular and organism level worked. The rediscovery of the principles of transmission genetics set the stage for some of the most important discoveries in the history of mankind. Over the course of the century, genetics began to permeate all of the life sciences. Scientists began to predict the probability of a disease using statistics and **pedigree** analyses.

All diseases are the result of biochemical problems at the cellular level, and diseases of the digestive tract are no exception. Over the last thirty years, the advances in cell biology and genetics have revolutionized medicine. In cell biology, discoveries of cell signaling and development have unveiled some of the complex chemical signals that occur between the cells of the body. In genetics, the revealing of the structure of DNA in 1953 by James Watson (1928–) and Francis Crick (1916–), the door was opened for the development of techniques to manipulate DNA in the laboratory. The completion of the **Human Genome Project** in 2003 will eventually enable the isolation of the genetic mechanisms of any disease, including those of the gastrointestinal system.

Advances in technology, specifically those to identify and treat diseases, also developed rapidly over the course of the twentieth century. At the end of the nineteenth century, physicians were still primarily involved in the treatment of symptoms, mostly due to a lack of techniques to discover the root causes of the ailment. During the twentieth century, procedures such as **magnetic resonance imaging** (MRI), **computerized axial tomography (CAT) scans, x-rays**, and **ultrasounds** became accessible to most doctors in the western world. Ultrasound analysis was initially developed by Ian Donald (1910–1987) in the 1950s to diagnose abdominal tumors; only later did it become associated with examining the health of the fetus prior to birth. With the availability of these diagnostic techniques, physicians are no longer confined to examining the appearance of the tongue or body fluids to determine the cause of a disease. Instead, they can peer directly into the interior of the human body and observe the action of the digestive organs.

A new area of research called computer science originated in the twentieth century. Starting from a simple computer with limited abilities to conduct mathematical calculations in the 1940s, computers had developed into their now recognizable form by the 1970s. In the last three decades of the twentieth century, computers began to have a tremendous influence on the life sciences. The invention of super-computers with incredible capacities for data processing has enabled scientists to quickly perform calculations and analyses that previously were either impossible or prohibitably time-consuming. The expansion of the Internet from a military-communication device to its present state as an information portal has placed invaluable re-

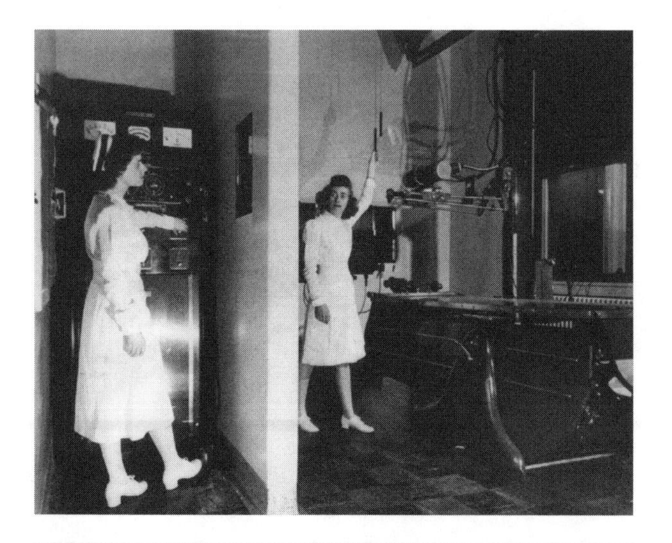

X-ray machine at the George Washington University Hospital, ca. 1950. © National Library of Medicine.

sources at the fingertips of scientists. Online databases, document sharing, and practically instantaneous communication have all played an important role in the rapid scientific advances in the fields of medicine, physiology, and genetics (to name a few) in the past thirty years.

While the study of inheritance enabled patterns and mechanisms to be determined, and technology provided the diagnostic tool, pharmaceutical drugs gave physicians the ability to treat diseases of the digestive system. As with other areas of twentieth century medicine, the use of drugs changed over the course of the century from initially serving to alleviate the symptoms of the disease, to the latest advances in treating the causes of the illness. One example is the treatment of acid indigestion. Initially treated with antacids, which removed the excess acid from the stomach but did little to address the cause, acid indigestion is now treated with designer drugs (see Chapter 9) that inhibit the action of proton pumps in the stomach, thus decreasing the ability of the cells to produce hydrochloric acid.

This does not mean that other areas of the sciences did not make contributions to the study of the digestive system. The determination that a microbe was responsible for certain forms of stomach ulcers (see Chapter 9) and the characterization of pathogenic variants of intestinal flora (Chapter 10) are the result of the work of medical microbiologists. Advances in the field of radiology allowed scientists to examine the internal workings of the intestinal tract using radioactive compounds such as barium. Chemists, specifically the biochemists, have frequently used digestive enzymes as model systems to explore the basics of biologically important reactions.

DISCOVERY OF THE VITAMINS

One of the more important areas of discovery during the twentieth century was study of the metabolic processing of nutrients. Researchers around the world were investigating carbohydrate metabolism, the biogenesis of cholesterol, and the energy-releasing pathways. A number of these were important enough to be awarded Nobel Prizes in either chemistry or physiology or medicine (see "Nobel Prizes"). However, some of the most important advances in the study of nutrient biochemistry occurred in the discovery and classification of the vitamins. The majority of these discoveries are important for the fact that they link dietary deficiencies to a specific disease, thus effectively beginning the study of nutrition as a science.

The term vitamin was introduced in 1912 by the Polish biochemist Casimir Funk (1884–1967). For some time previously, scientists—in experiments on both animals and humans—had noticed that a diet of protein, carbohydrates, and fats was not sufficient to maintain normal health. For example, in the late nineteenth century, two seemingly unrelated diseases, beriberi in humans and polyneuritis in chickens, were being associated with

Nobel Prizes

One of the greatest recognitions that may be bestowed upon a scientist is the Nobel Prize. Awarded to "those who, during the preceding year, shall have conferred the greatest benefit to mankind," the Nobel Prizes are given on behalf of the estate of Alfred Nobel (1833–1896). Nobel was a nineteenth century inventor and chemist who was responsible for stabilizing the explosive compound nitroglycerine by combining it with diatomite. The result was dynamite. Nobel became rich from his invention, but despaired at being called the "merchant of death." The Nobel Prizes, awarded annually since 1901, with the exception of during some periods of international conflict, are Alfred Nobel's attempt to leave something beneficial for mankind.

In the twentieth century, Nobel Prizes that are associated with the digestive system are usually awarded in the category of Physiology or Medicine. However, some early studies of enzyme function that relate to digestive processes were awarded Nobel Prizes in Chemistry. The Nobel Prizes that relate in some manner to the digestive system are listed below.

1904	Ivan Pavlov (Russian): The discovery of the interaction between the nervous and digestive systems.
1923	Frederick Banting (Canadian) and John MaCleod (Scottish): Isolation of insulin from the pancreas.
1928	Adolf Windaus (German): The relationship between sterols and vitamin D.
1929	Frederick Hopkins (English): The relationship between vitamins and human diseases such as scurvy and rickets.
	Christian Eijkman (Dutch): The relationship between human diseases, such as beri-beri, and dietary deficiencies.
1937	Walter Haworth (English): Investigations of carbohydrate metabolism and research on the synthesis of vitamin C.
	Paul Karrer (Russian): The relationship of vitamin A to carotenoids and the isolation of vitamin B_2 (riboflavin).
	Albert Szent-Gyorgyi von Nagyrapolt (Hungarian): Carbohydrate metabolism.
1938	Richard Kuhn (Austrian): Research on B vitamins.
1943	Edward Doisy (American): Structure and synthesis of vitamin K.
	Carl Dam (Danish): Vitamin deficiencies and the role of vitamin K.
1947	Carl Cori (Czech) and Gerty Cori (Czech): The synthesis of glycogen.
	Bernardo Houssay (Argentinan): The endocrine control of carbohydrate metabolism.
1953	Fritz Lipman (German): The isolation of acetyl-coenzyme A, an important intermediate in nutrient metabolism.
	Hans Krebs (German): Discovery of the citric acid cycle.

1955	Axel Theodor Theorell (Swedish): Action of enzymes, including riboflavin.
1958	Frederick Sanger (English): The molecular structure of insulin.
1964	Konrad Bloch (German): The structure of cholesterol.
	Feodor Lynen (German): The synthesis of cholesterol in living cells.
1981	Roald Hoffman (Polish): Vitamin B_{12} synthesis.
1985	Michael Brown (American) and Joseph Goldstein (American): Cholesterol metabolism in humans and the relationship to lipoprotein receptors.
2002	John Fenn (American), Kiochi Tanaka (Japanese), and Kurt Wüthrich (German): The use of mass spectrometric analysis in examining biological macromolecules.

Note: List includes only the winners of the Nobel Prize in Chemistry or the Nobel Prize in Physiology or Medicine whose discoveries were related to major discoveries or advances in the study of digestion, enzymes, or nutrition.

diets that relied heavily on polished, or white, rice. Some, including the 1929 winner of the Nobel Prize, Christian Eijkman (1858–1930), initially believed that the diseases were due either to toxins introduced during the processing of the rice, or to bacteria. But others, such as Eijkman's partner Gerrit Grijns (1865–1944), suspected that dietary deficiencies were involved since diets rich in brown (unpolished) rice rarely produced problems. Using chickens as a model, Eijkman and Grijns performed an experiment in which some chickens were exclusively fed white rice, while others were sustained on a diet of brown rice. As expected, the birds that were fed white rice fell ill. Since polyneuritis and beriberi are similar diseases, Eijkman was led to believe that diet was also a cause in beriberi. Eijkman surveyed a number of prisons and established that as the case was with polyneuritis, there was a direct correlation between diet and the disease, since inmates fed brown rice rarely developed beriberi. From these and other ongoing studies of pellagra and scurvy, Funk developed the concept of a vitamin. He believed that these nutrients were derived from ammonia compounds called amines. From the word amine he developed the word vitamine (an abbreviation of "vital-amine"). Once it was discovered that not all vitamins are related to the amines, the final *e* in the word was dropped, and the class of nutrients is now simply called vitamins.

Once it had become apparent that vitamins were an important component of diet, the next logical step was to isolate the compounds from foods so that they could be used to treat diseases caused by dietary deficiencies. By the early twentieth century a number of researchers were actively working

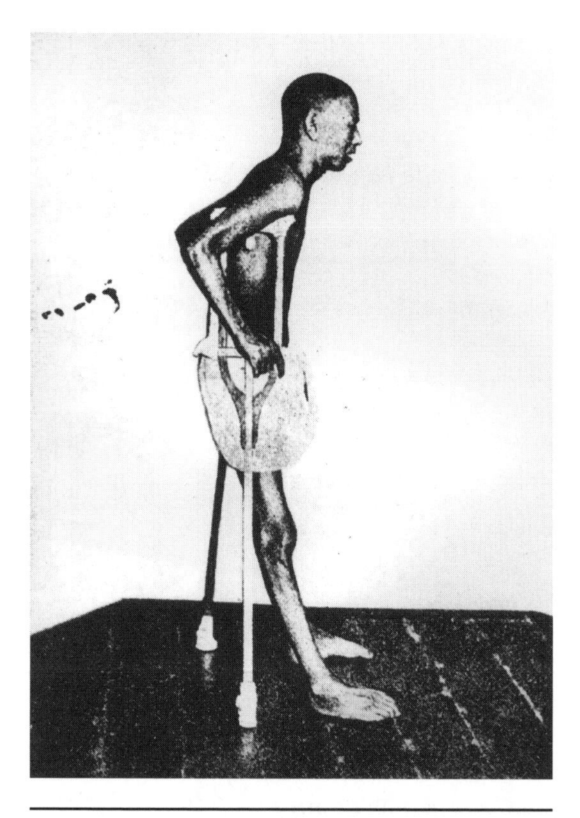

An inhabitant of the Dutch East Indies suffering from severe beriberi (vitamin B_1 deficiency). © National Library of Medicine.

on this project. One of the first successful isolations of the nutrient that prevented beriberi was done by a team of researchers in Japan in 1912. Several years later Funk performed a similar isolation. In 1913, an American biochemical team headed by Elmer McCollum (1879–1967) isolated a fat-soluble compound from butter that appeared to have all of the qualities of vitamins. McCollum named this nutrient vitamin A, and assigned the vitamin responsible for preventing beriberi vitamin B.

The naming of vitamins gets a little more convoluted after this time. The antiscurvy factor, ascorbic acid, was assigned the letter C (vitamin C). However, ascorbic acid was really the first vitamin whose existence was suspected by the medical community (see "Scurvy and Vitamin C"). The chemical structure of vitamin C was independently determined in 1932 by Charles King (1896–1988) and Albert von Nagyrapolt Szent-Györgyi (1893–1986). A synthetic form of vitamin C was first manufactured the next year by both Walter Haworth (1883–1950) and Tadeus Reichstein (1897–1996). Haworth received the Nobel Prize in Chemistry in 1937 for his work on the biochemistry of carbohydrates. Reichstein later received the 1950 Nobel Prize in Chemistry for his work on hormones.

The next vitamin was vitamin D. The existence of vitamin D, another of the fat-soluble vitamins (see Chapter 1), was first alluded to by the English scientist Edward Mellanby (1884–1955) around 1921. Mellanby was working on the causes of a disease of the skeletal system called **rickets**, when he discovered that diets containing animal fats, or oils from animals such as cod liver oil, prevented the disease. Oddly enough, people exposed to sunlight, but lacking vitamin D in the diet, also demonstrated a reduced level of rickets. As noted in Chapter 1, vitamin D is an important nutrient of the digestive system in that it signals the intestine to absorb calcium. The chemical structure of vitamin D was first discovered by several researchers, including the German scientist Adolf Windaus, who would later receive the Nobel Prize in Chemistry for his work with sterols. It is now known that vi-

Scurvy and Vitamin C

Since early times, populations that have been isolated from fresh fruits and vegetables for extended periods of time have been known to be exceptionally susceptible to **scurvy**. Colonies located in inhospitable regions, cities under siege during war, and sailors on extended sea voyages have all been afflicted by this disease. One of the first signs of scurvy are bleeding gums and loose teeth. If left untreated, the patient gradually loses immune system function, become progressively weaker and eventually dies.

All of the great European explorers of the of the fifteenth and sixteenth centuries documented cases of scurvy. In many cases over half a ship's crew perished from scurvy during the course of a single expedition. Diets on these voyages were especially bad, with rice and salted meats comprising the majority of the food. Fresh fruits and vegetables were virtually nonexistent. The great sea powers of the sixteenth and seventeenth centuries, notably Spain, England, and France became especially concerned since their international interests were based on their naval power, and if a crew was afflicted with scurvy is was possible to lose battles to a vastly undersized enemy.

In the eighteenth century, the Scottish surgeon James Lind (1716–1791) conducted a series of experiments to find a means of preventing scurvy. He chose twelve patients, paired them into groups and then gave each group a different remedy. Fresh fruit, vinegar, garlic, and sea water were all used. The sailors that were administered the fresh fruit recovered rapidly. It is important to note that Lind did not identify scurvy as a dietary deficiency disease, but rather thought that it was associated with moist sea air, and that the fruit was preventing the action of the air. However, based on Lind's recommendations, the Royal Navy adopted a practice of issuing a daily ration of lime juice to its sailors. Scurvy practically disappeared in the British Navy, and their sailors became known as "limeys."

The link between vitamin C and scurvy was established over 100 years later, mostly due to the work of the Polish-born biochemist Casimir Funk (1884–1967). From his previous work on vitamins, he recognized that a nutrient factor present in the fruits was preventing scurvy. He was also one of the first to determine that this factor was destroyed by heat processing. Vitamin C is also commonly called ascorbic ("antiscurvy") acid.

---

tamin D may be synthesized through a biogenic pathway that converts the sterol 7-dehydrocholesterol to vitamin D through the interaction of the skin, kidneys, and liver. (The similarities of vitamin D and 7-dehydrocholesterol are shown in Figure 7.1.)

Vitamin E, an antioxidant, fat-soluble vitamin (see Chapter 1) that belongs to a class of compounds called the *tocopherols*, was first isolated in 1922 by the American scientists Herbert McLean Evans (1905–1983) and Katherine Scott Bishop (1889–1976). Unlike the discoveries of the previous vitamins, the isolation of vitamin E was not initiated as an investigation into

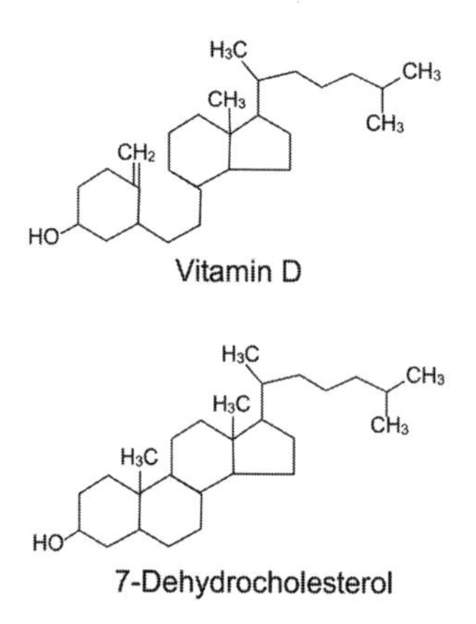

Figure 7.1. The similarities between vitamin D and 7-dehydrocholesterol, a cholesterol naturally present in the body. Vitamin D may be generated from 7-dehydrocholesterol by the action of the skin, kidneys, and liver.

any specific disease. In fact, the exact role of vitamin E in the body is still under investigation.

Unfortunately, the vitamin B that was being studied in the early twentieth century was actually a combination of several different chemicals. The anti-beriberi nutrient is actually thiamin. The structure of a water-soluble vitamin with a similar structure, but lacking the anti-beriberi properties of thiamin, was determined in the early 1930s through the efforts of a number of scientists, most notably the Austrian chemist Richard Kuhn (1900–1967). Kuhn received the 1938 Nobel Prize in chemistry for his work on the B vitamins. Originally called vitamin F, it was later renamed vitamin B_2, with the name vitamin B_1 reserved for thiamin. Vitamin B_2 is commonly called riboflavin. The name is derived from the Latin word for "yellow," since riboflavin has a yellow color. It was artificially synthesized in the 1930s by the Swiss chemist Paul Karrer (1889–1971), who received the 1937 Nobel Prize in chemistry for his efforts.

Another of the B vitamins of interest to the twentieth century is niacin. Niacin belongs to the B vitamin complex, and thus is a water-soluble vitamin. Niacin deficiencies cause the disease *pellagra*, which is characterized by dry skin, diarrhea, and an inflamed tongue. In many cases, patients experience various levels of mental disease. It is a common disease of poor areas of the world where the diet lacks milk or eggs. During the 1920s the American physician Joseph Goldberger (1874–1929) studied the diets of prison populations where pellagra was present. Like many other dietary diseases, pellagra was initially believed to be caused by bacteria, until Goldberger noticed that the staff of the prisons rarely were afflicted with pellagra, only the inmates. When milk products and eggs were added to the inmates' diet, the cases of pellagra plummeted. Goldberger initially named this nutrient the *P-P factor*, for pellagra-preventative. In 1937, the American scientist Conrad Elvehjem (1901–1962) demonstrated that P-P factor belonged to a class of chemicals called the nicotinamides. To avoid the misconception that nicotine, the active compound of tobacco, was a vitamin, Elvehjem named the chemical niacin.

The "discovery" of other B vitamins followed, as did the sequential numbering. However, many of these discoveries turned out to be false, and thus there are no B vitamins numbered from 3 to 5. A group of the B vitamins, namely pantothenic acid, folic acid, and biotin, were discovered in the 1940s as physicians were investigating the role of intestinal bacteria in human digestion.

Of special historical interest is vitamin B_{12}. It is sometimes also called an antianemic factor, due to the nature of its discovery. In the mid-1900s a disease called pernicious anemia was first diagnosed. Patients suffering from this disease failed to produce adequate red blood cells and eventually died. For the next seventy years physicians failed to make any progress on the cause of the disease, until experiments in the 1920 by George Whipple (1878–1976), an American scientist. Whipple did not have a direct interest in pernicious anemia, but rather was studying bile pigments found in the blood. To gather enough blood for his experiments, Whipple bled dogs until they were anemic. He experimented with a number of diets to determine which would replace the lost hemoglobin fastest, and discovered that a diet rich in liver worked best.

Several years later two scientists, William Murphy (1892–1936) and George Minot (1885–1950) decided to use a diet rich in liver for patients with pernicious anemia. They found that if the patients were fed large quantities of liver that they rapidly recovered from their illness. However, the quantities had to be very large, and the liver had to be raw, or at most lightly cooked. They had found a cure for the disease, although they had not isolated the dietary factor. For their work, Whipple, Murphy, and Minot received the 1934 Nobel Prize in physiology or medicine.

Around 1929, the American physician William Castle (1897–1990) discovered that the composition of gastric juice in patients suffering from pernicious anemia differed from that of the normal population. He believed that there were two parts to the problem. The first was the availability of a nutrient that was necessary in the manufacture of new blood cells. He called this an *extrinsic factor*, and it is vitamin B_{12}. Second, he believed that the stomach lining of normal people secreted a second compound in the gastric juice that allowed for the absorption of the extrinsic factor. This he called an *intrinsic factor*. In 1954, it was determined that this intrinsic factor was a lipoprotein. Thus Castle had solved the riddle of why large amounts of liver were needed by patients with pernicious anemia. Without the intrinsic factor, the extrinsic factor could not be absorbed in sufficient quantities while the food was in the stomach. In 1948, another interesting fact regarding vitamin B_{12} was discovered: it is the only chemical in the body that contains the mineral cobalt.

Finally, there is vitamin K. This vitamin is a fat-soluble vitamin, and in healthy people is manufactured by intestinal bacteria. It was first discovered in 1934 by the Danish chemist Carl Dam (1895–1976) while working on poultry. Dam discovered that when this nutrient was missing from the diet, the chickens had difficulty with blood coagulation. The letter K denotes the German spelling *koagulation*. In 1939, the American scientist Edward Doisy (1893–1986) determined the structure of the vitamin. For their work, Doisy and Dam shared the 1943 Nobel Prize in Physiology or Medicine.

SIGNALING IN THE DIGESTIVE SYSTEM

Unlike earlier centuries, when research on the digestive system was being performed by a limited number of scientists and physicians at a few institutions, the twentieth century was a time when the science of physiology proliferated significantly. The nineteenth-century concept of physiology was replaced by a variety of sub-disciplines of life science in the twentieth century, mostly due to specialization of the scientific community. Physiology at the cellular level, an area unknown to nineteenth-century scientists, became an important aspect of twentieth-century biochemical studies. Other researchers focused on the physiology of select organ systems or the interaction of several systems. Still, physiologists continued to use diseases as the basis for determining the workings of the human body.

One of the more interesting discoveries of physiology involving the interaction of two body systems was made by the Russian physiologist Ivan Pavlov (1849–1936). Pavlov was interested in the reflex actions involving the nervous system. He used the response of the digestive system to demonstrate that living organisms are influenced by their environment. In his experiments, Pavlov placed food into a canine's mouth and observed that there was an increase in gastric juice secretion by the stomach (see Chapter 2 for a discussion of this reflex). He then exposed the dogs to the aroma of food, and again the stomach increased gastric juice production. If a bell was rung while the dogs were being given food, then the ringing of the bell at a later time would result in increased gastric juice production, even if no food was being offered to the dog. Although Pavlov used these experiments to define reflex actions, they also clearly demonstrated a connection between the nervous system and digestive system.

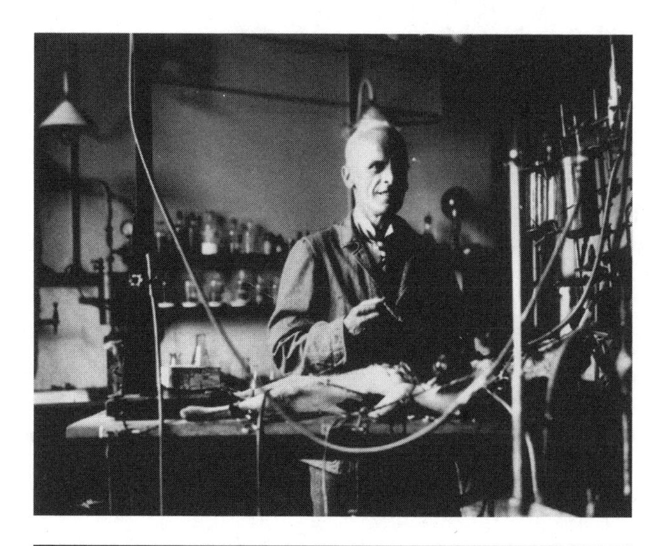

Ernest Starling at work in his laboratory. © National Library of Medicine.

The proof of the connection between these systems was provided in 1902 by two English scientists, Ernest Starling (1866–1927) and William Bayliss (1860–1924). It was known that the pancreas began secreting digestive enzymes as food entered into the stomach. Starling and Bayliss were interested in determining the mechanism by which the stomach signaled the pancreas. Pavlov's work suggested that there was a link with the nervous system. However, when Starling and Bayliss surgically cut the nerves to the pancreas it still secreted digestive enzymes into the small intestine when food was present in the stomach. Thus there had to be something more involved than Pavlov's reflex actions. In other experiments, Starling and Bayliss determined that it was a chemical signal that was released by the small intestine that was responsible for the pancreatic reaction. When hydrochloric acid is present in the duodenum, a chemical signal stimulated the release of pancreatic enzymes. Starling and Bayliss called this chemical secretin. Not only had they identified the first of the chemical signals in the digestive system, they had also discovered one of the first hormones (see the Endocrine System of this series volume for more information on hormones).

SURGERY

The twentieth century is called by some the "Golden Age" of surgery. Due to the invention of anesthetics and aseptic techniques in the nineteenth century, surgery was now a viable option for the treatment of many ailments, including some of the digestive system. During the twentieth century, appendectomies (the removal of the appendix) became a common surgical technique. This technique gained international attention when, in 1901, the appendix of Edward VII was removed just prior to his becoming king of England. Unfortunately, until the use of antibiotics became prevalent following World War II, the mortality rate due to secondary infections was very high (sometimes 50 percent) for patients of appendectomies.

The advent of new technologies and drugs to suppress the immune system meant that by the 1960s organ transplants were also common. While the majority of the early transplants involved the heart or kidneys, in 1967 the first liver transplant was made by physicians at the University of Colorado. It is now recognized that the redundant structure of the liver means that the patient does not need to receive a full liver since following a partial liver transplant the liver will regenerate in both the donor and recipient.

ANATOMY AND PHYSIOLOGY

The work of the nineteenth-century scientists and physicians on the anatomy and physiology of the digestive system. Although the debate on the role of the appendix as a digestive organ (see Chapter 3) continues to this day, the roles of the major organs were well established by the early twentieth century. The majority of discoveries in the twentieth century that related to anatomy and physiology were associated with the study of specific diseases and ailments of the digestive tract, and thus will be covered in more detail in the following chapters.

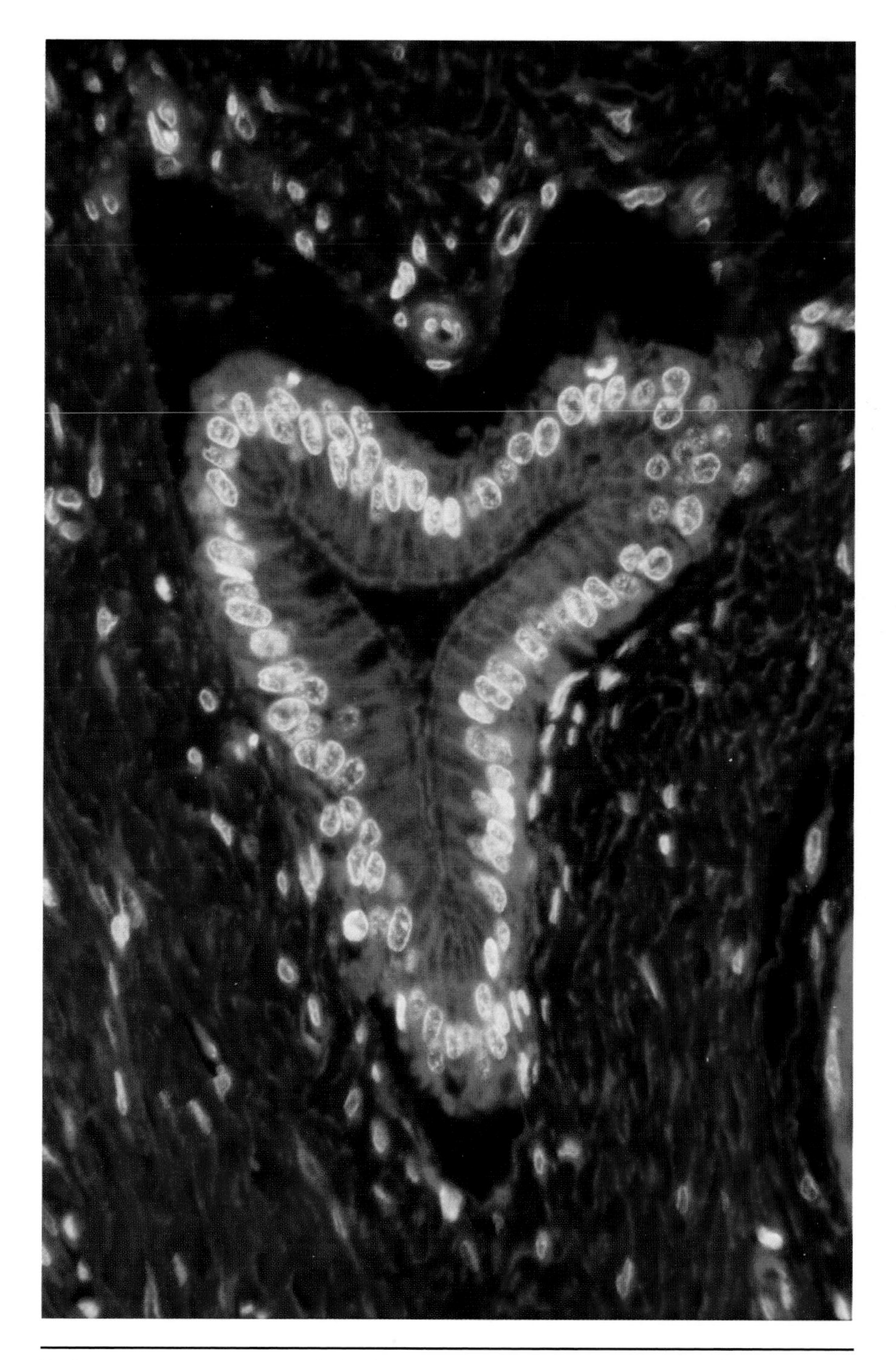

Epifluorescence microscopy of the human liver bile duct; cross-section. © Albert Tousson/Phototake—All rights reserved.

Scanning electron micrograph of mammalian tongue papillae and a taste bud, magnified 52×. © Dennis Kunkel/Phototake—All rights reserved.

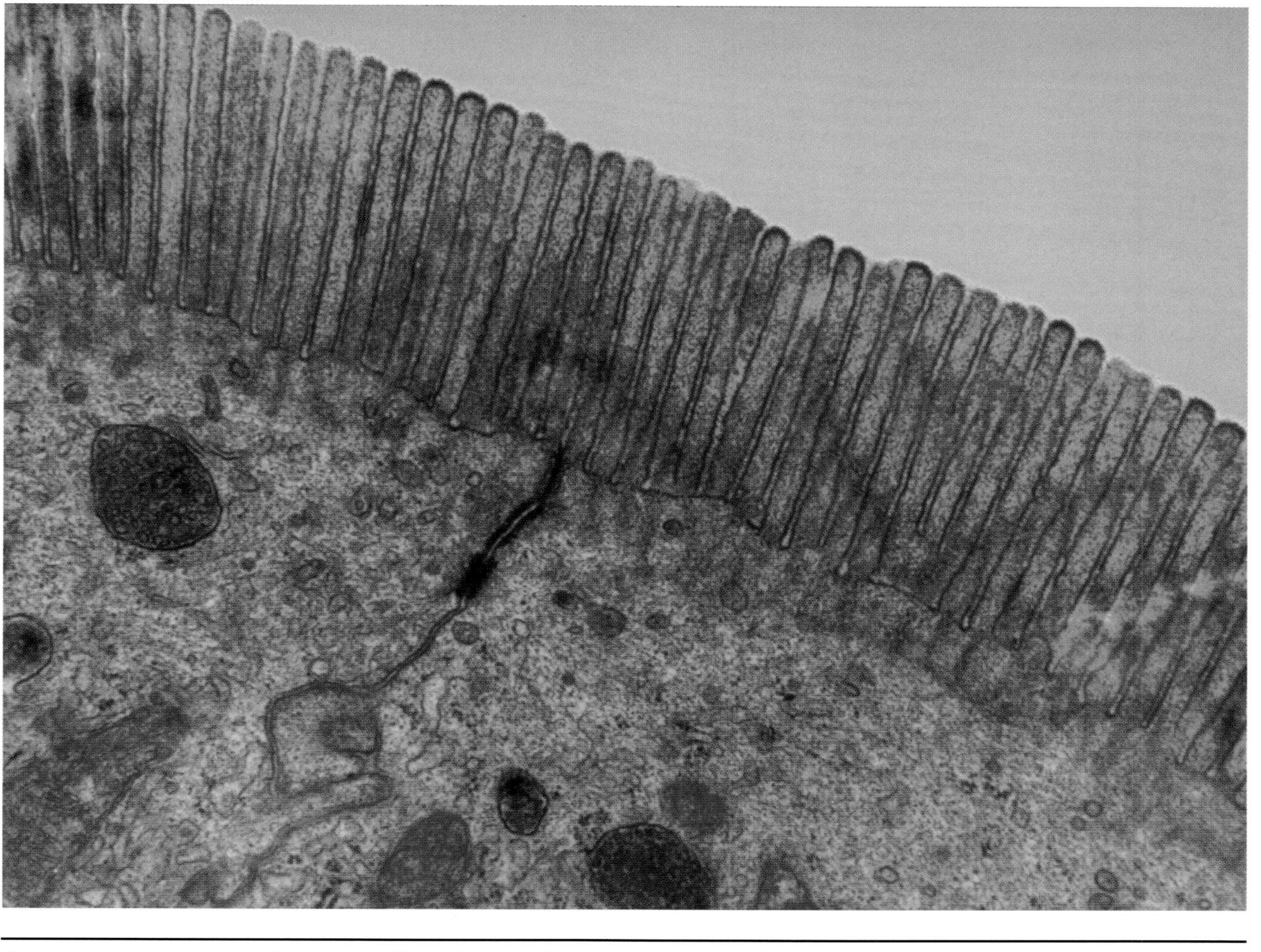

Electron micrograph of small intestine villus and microvilli, magnified 6,480×. © Dennis Kunkel/Phototake—All rights reserved.

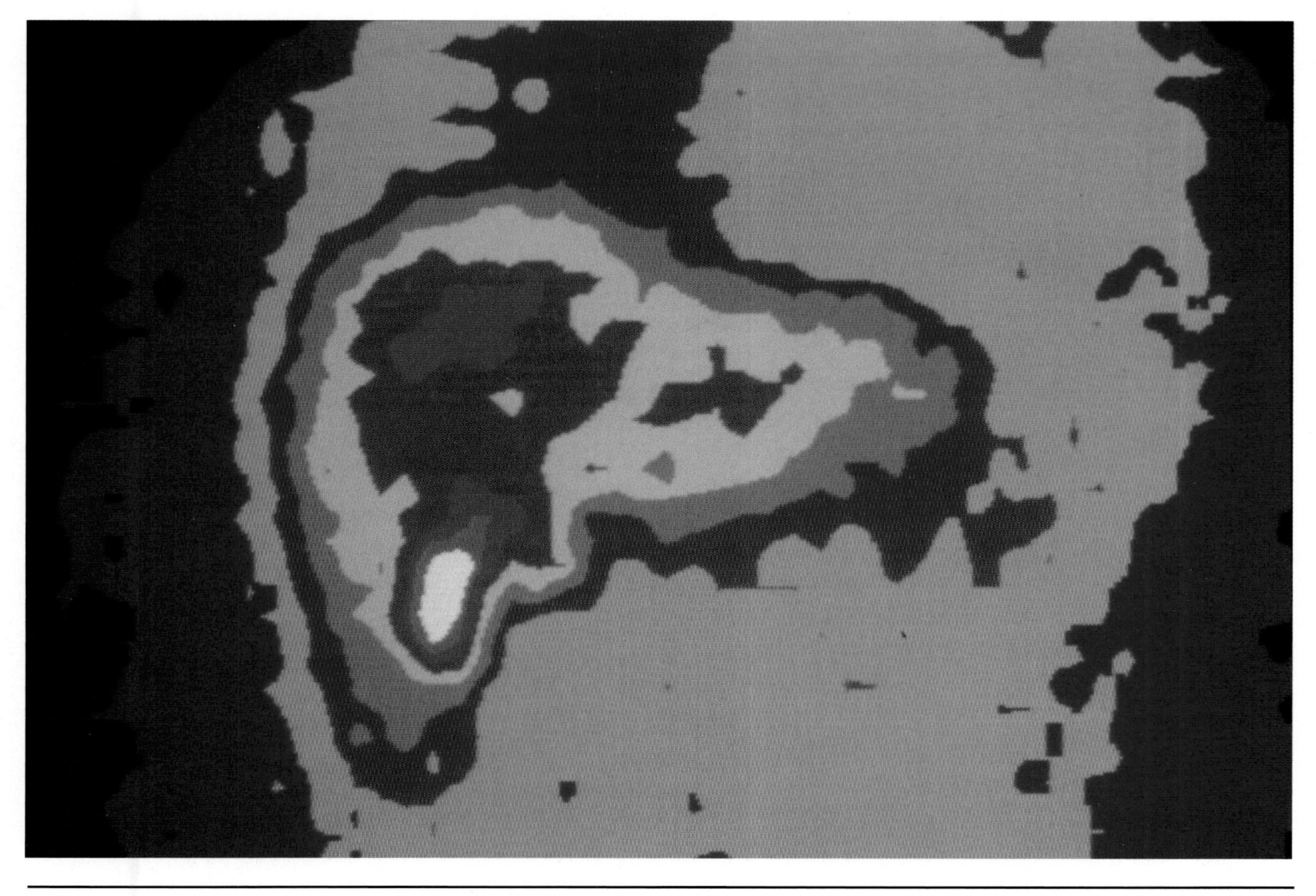

Nuclear medicine scan of a normal liver with gall bladder shown in the white area. © Collection CNRI/Phototake—All rights reserved.

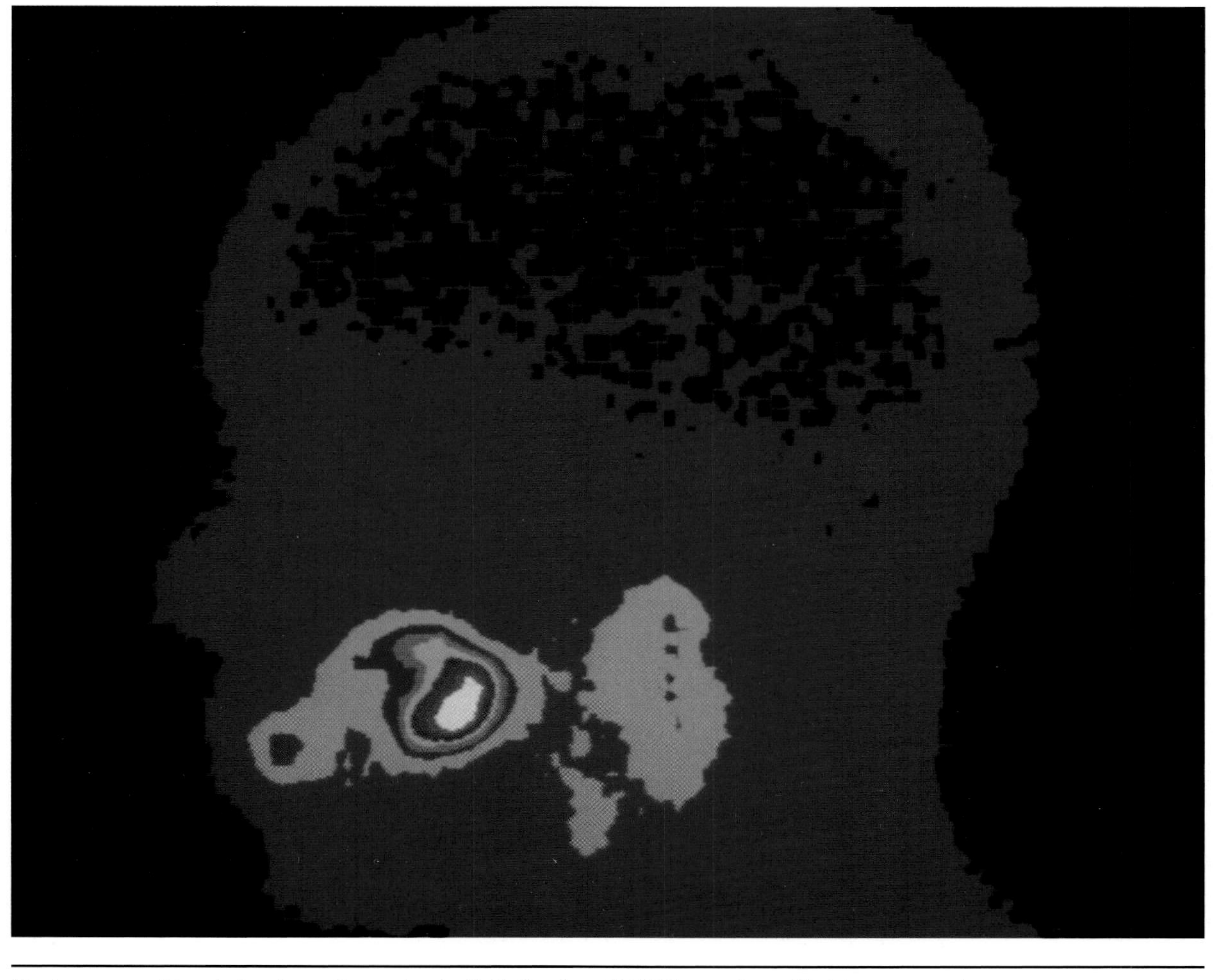

Nuclear scan of head showing salivary glands. © Collection CNRI/Phototake—All rights reserved.

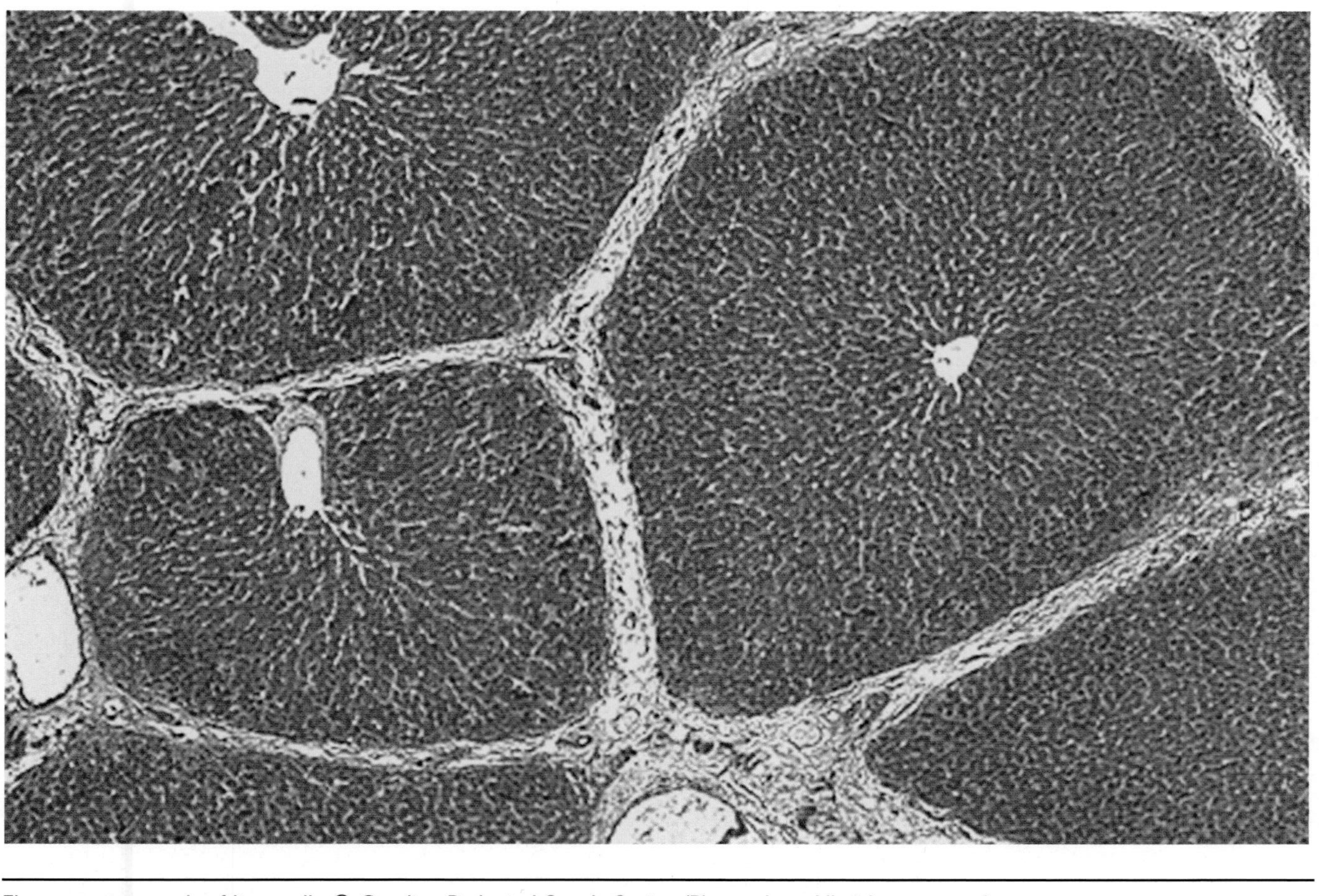

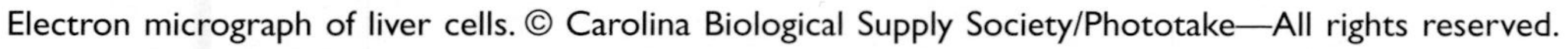

Electron micrograph of liver cells. © Carolina Biological Supply Society/Phototake—All rights reserved.

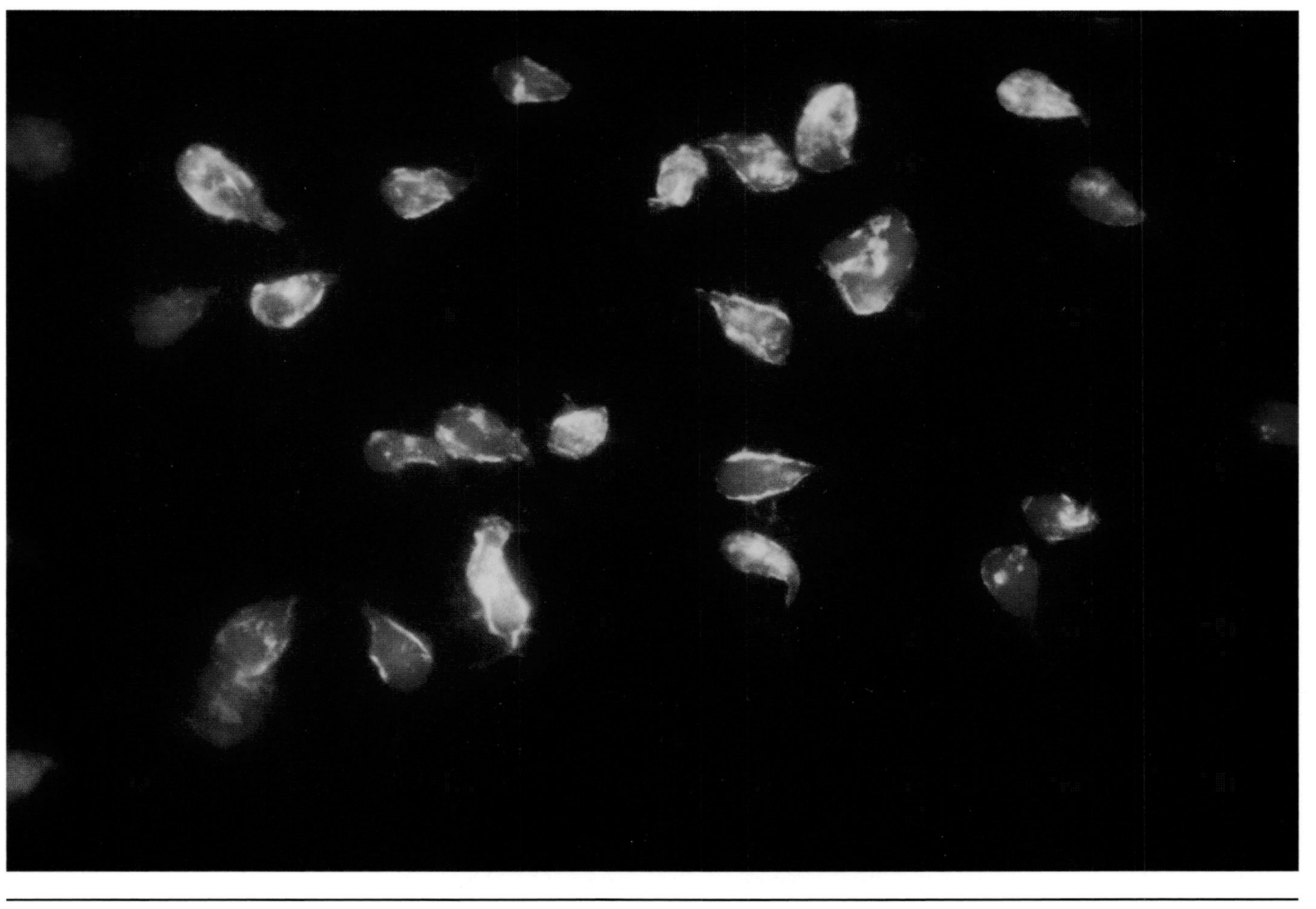

Salmonella bacteria in tetrathionate enrichment broth stained using direct FA staining technique, 1969. © Centers for Disease Control and Prevention.

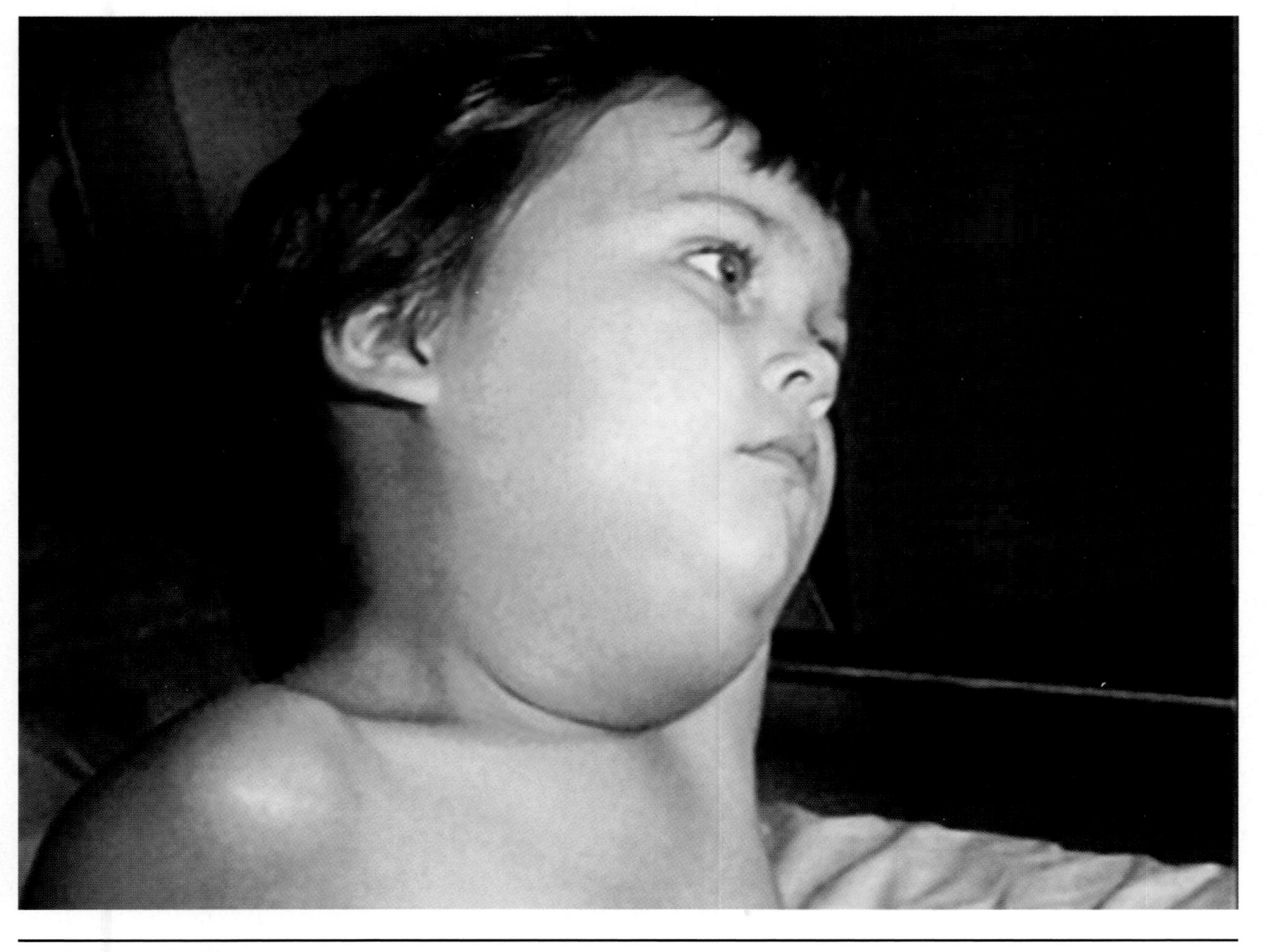

A child with mumps. © Centers for Disease Control and Prevention.

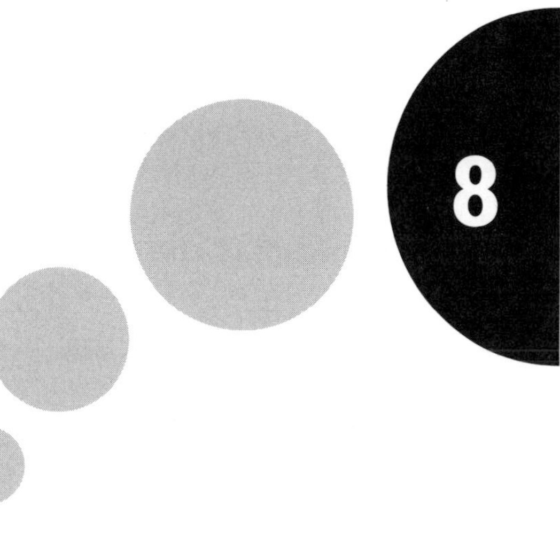

Diseases and Ailments of the Oral Cavity and Esophagus

The upper GI tract includes the oral cavity, esophagus, and stomach. This chapter will examine the ailments of the oral cavity and esophagus. The stomach will be covered in the next chapter. While the salivary glands, an accessory gland of the oral cavity, are technically associated with the upper GI, problems associated with their function will be discussed in Chapter 11. In addition, since cancer is common to all of the gastrointestinal glands, it will be covered separately in Chapter 12.

THE ORAL CAVITY

As the opening to the gastrointestinal system, the oral cavity represents much more than a place for mechanical digestion to be initiated. The health of the oral cavity gives a medical practitioner an insight into the diet of the individual, level of personal hygiene, and more importantly, the overall health of the remaining hidden sections of the gastrointestinal tract. Ailments of the stomach and lower GI frequently alter the chemistry of the oral cavity, resulting in sores, problems with the teeth, bad breath (*halitosis*), and other symptoms. Ensuring the health of the oral cavity also serves not only to extend the lifespan of the teeth, but also protects the remaining organs of the digestive system from infection.

For organizational reasons, the ailments of this component of the digestive system will be characterized as belonging to one of two general classes. First, we will examine problems, diseases and ailments of the overall oral cavity. This will be followed in the next sections by a more detailed look at ailments and diseases specific to the teeth and gums (gingivae).

Ailments of the Oral Cavity

SORES OF THE ORAL CAVITY

Although they may seem to be related, canker sores and fever blisters actually represent two completely different types of oral diseases. Canker sores, or *aphthous stomatitis*, occur within the interior of the oral cavity, while fever blisters occur on the lips. Furthermore, the cause of most fever blisters is known to be viral, while canker sores may be the result of a wide number of factors.

Canker sores get their name from the French word *canker*, which means a spreading sore or ulcer. The term *cancer* is derived from the same term. These sores occur along the mucus membranes of the tongue, cheeks, and lips. These sores usually originate as small (⅛ inch to ½ inch wide) swellings. Within twenty-four hours, these swellings burst and are covered by a yellow or white covering. During this time they are very sensitive to acidic or spicy foods and frequently cause considerable discomfort. The sores usually heal completely within one to two weeks. Although canker sores may be very painful, they very rarely cause additional medical problems unless they become infected.

No one is exactly sure as to what causes canker sores in an individual. A number of researchers believe that canker sores may be the result of an **autoimmune response** on behalf of the body against cells lining the interior of the oral cavity. In an autoimmune response, the body mistakenly identifies types of its own cells as pathogens, or invaders, and begins an immune response against those cells (see the Lymphatic System volume of this series for additional information). Other researchers believe that the sores are the result of food allergies or nutritional deficiencies. Canker sores may also be an indication of other gastrointestinal problems. While the cause of canker sores may be elusive, scientists are reasonably certain that they are not caused by bacterial or viral infections, although the normal flora of bacteria in the oral cavity can infect a sore once it has ruptured. This does mean that canker sores are not contagious and can't be transmitted by kissing or sharing utensils.

Unfortunately, since there is individual variation in the cause of canker sores, there is no universal treatment. Medication may be used to reduce the pain, but it appears that the one- to two-week lifespan of a sore must be completed naturally. Antiseptic rinses of the oral cavity may reduce the potential for bacterial infection. Avoidance of certain foods, such as those high in spices, may also reduce discomfort. It is recommended that those individuals who routinely suffer from canker sores contact their physician and discuss allergy screening.

Fever blisters resemble canker sores in many ways, but occur on the outside of the mouth, usually in the area of the lips. Fever blisters are contagious since they are caused by a virus called *herpes simplex*. The herpes virus (from the Greek word *herpo*, meaning "to creep") is known to exist in two forms, herpes simplex-1 and herpes simplex-2. While both forms may cause fever blisters, herpes simplex-2 is usually associated with genital herpes (see Reproductive System volume of this series). Herpes simplex is a lysogenic form of virus, meaning that once it infects the body it integrates into the cells and remains dormant until it receives a signal to activate. Scientists know that the herpes simplex-1 remains dormant in the nerve cells of the face and then moves to the tissues surrounding the mouth once activated. Unfortunately, the stimuli and chemical signals that activate the virus are unknown. It appears that stress, sunlight, illness, or injuries may activate the virus, but in other cases the virus may activate without any known stimulus.

There are no known cures for herpes simplex and once a person is infected with the herpes virus they remain infected for their entire life. It is believed that upwards of 80 percent of the United States population is infected with the virus. Research is underway to develop medicines that inhibit the outbreak cycle or vaccines that make individuals resistant to infections. Some research has suggested that the number and severity of the outbreaks decreases with age, but it is unclear as to why this occurs. In addition, some of the new antiviral medications, such as Acyclovir topical, show promise.

THRUSH

Thrush is the common name for a condition called *candidiasis*. The disease may infect both the oral cavity and esophagus. Candidiasis infections are also common in women as vaginal yeast infections. Oral candidiasis is an infection caused by members of the *Candida* **genus** of fungi. The most likely culprit is the fungus *Candida albicans*, a common fungus that is frequently found living on the moist membranes of humans. In people with healthy immune systems, the natural levels of the fungus are kept under control. However, if the immune system is not functioning correctly, or the chemistry of the oral cavity is altered significantly, then the fungus may quickly invade the surrounding tissues. The disease usually affects infants whose immune system is not yet completely functional, although anyone that has been on extended use of antibiotics, or has a compromised immune system due to HIV or chemotherapy, is at risk. Medical conditions that produce dry mouth, such as Sjögren's syndrome or xerostomia (see Chapter 11) may enhance the chance of developing thrush.

Candidiasis is characterized by creamy white patches within the mouth that is often compared in appearance to cottage cheese. The membranes of the mouth may also be red or swollen. Since, as organisms fungi digest food outside the body and then absorb it into the fungal cells (a process called extracellular digestion), candidiasis may cause a foul taste in the mouth and bad breath, or halitosis. The disease is easily treated with anti-fungal medications.

CLEFT PALATE

Cleft palate and cleft lip are birth defects and as such represent a problem with the embryonic development of the oral cavity. They are the most common of all birth defects and occur in about 1 out of every 700 births.

The roof of the mouth is formed by two structures, called palates. The hard palate is a bony structure that separates the nasal and oral cavities. The soft palate, located toward the rear of the oral cavity, is comprised of boneless tissue. During the development of the fetus (around ten weeks) the palates are formed from tissue that originates at the jaw bones and grows inward. These tissues unite to form the hard and soft palates. However, if this process is interrupted and the tissues fail to unite, the result is a cleft palate. The reason why the palates do not fit correctly is not completely understood, but it is most likely due to unfavorable environmental conditions during embryonic development. For example, smoking and alcohol use by the mother during pregnancy increases the chances of cleft palate, as does a deficiency in the B vitamin folic acid. A cleft palate can interfere with speaking, breathing, and eating if not corrected. In some cases a cleft palate can interfere with the normal positioning and development of the teeth. A cleft lip is similar to a cleft palate, but is more noticeable due to the characteristic appearance of the individual.

Recently researchers at the University of Iowa have identified a gene, called IRF6, on chromosome 1 that may be responsible for a genetic version of cleft palate called *Van der Woude syndrome*. This syndrome is an *autosomal dominant disease*, meaning that only one copy of the defective gene is necessary for the syndrome to develop and there is no difference in the rate of occurrence between males and females. As with cleft palate, Van der Woude's syndrome may present itself in a variety of ways, from minor cleft lips or missing teeth to fully cleft palates. Van der Woude syndrome is responsible for only 2 percent of all cases of cleft palate, but the identification of the genetic mechanism should shed light on the embryonic process that is disrupted resulting in cleft palate.

The degree that the palate is divided varies by individual. In one condition, called submucous cleft palate, either the soft or hard palate may be covered by a mucous membrane and only the underlying tissue remains divided. In some of these cases the individual may not be aware of a cleft palate until speech or teeth problems present themselves. Cleft palate may also be a sign of a more systemic disease called **velocardiofacial syndrome**.

Cleft palate and cleft lip are both correctable by surgery. The timing of the operation depends on the severity of the case and the interference with development, but in most cases, surgery is performed between three and twelve months of age. Individuals with cleft palate frequently have to undergo speech therapy following surgery.

Medical Problems of the Tongue

The tongue is technically a part of the muscular system, since it is comprised of skeletal muscle tissue covered by a mucous membrane. However, since the tongue plays a central role in the initial processing of food, the ailments of the tongue should be mentioned in this volume.

With the exception of cancer of the tongue (see Chapter 12), most problems associated with the tongue are not serious but rather indicate either a problem with the overall digestive system, poor oral hygiene, or malnutrition. For this reason the tongue has historically, and usually very accurately, been used as an indicator of gastrointestinal health.

Halitosis, or bad breath, is an indication of poor oral health. Like the other tissues of the mouth, the tongue is susceptible to bacterial growth, especially when the chemical conditions of the mouth are altered to favor bacterial or fungal growth. Most dentists now recommend that people of all ages brush their tongue the same way that they would treat the gums or teeth. This action alone may significantly reduce the incidence of halitosis.

The papillae of the tongue (see Chapter 2) normally give the muscle a rough appearance. The papillae themselves look like small whitish-pink raised dots along the surface of the tongue. If the tongue appears to be furry or to have black hairs on it, then this is a good indication of poor oral hygiene. The furry appearance is typically due to the accumulation of dead cells on the surface of the tongue. If these dead cells are colonized by oral bacteria, they may turn black and give the tongue a hairy appearance. The use of tobacco or long-term antibiotics may also cause these symptoms. A physician should be consulted if the condition does not readily clear up with adequate oral hygiene.

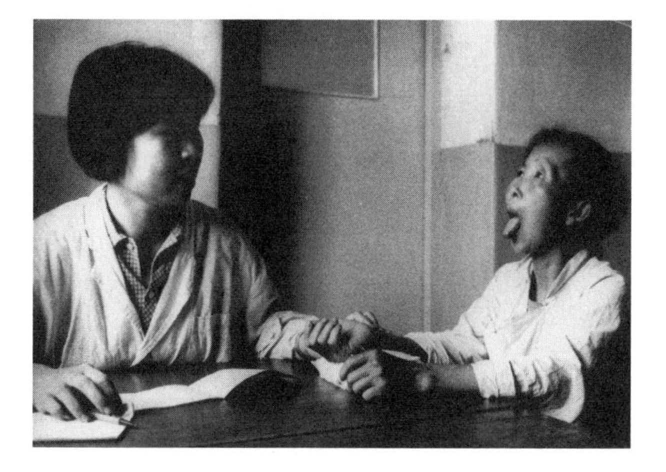

A doctor examines a patient's tongue, China. © WHO photograph by D. Henrioud/National Library of Medicine.

As mentioned, the appearance of the tongue may change due to malnutrition. For example, a smooth tongue most often represents a deficiency in the B vitamins. Typically this is associated with either riboflavin, vitamin B_{12} or folic acid. A smooth tongue can also be the result of an iron deficiency in the body. In all cases, a person should consult a physician before attempting to use supplements to correct the situation.

Diseases of the Teeth and Gums

GINGIVITIS AND PERIODONTAL DISEASE

The gums, or gingivae, comprise the tissue that are responsible for protecting the delicate roots of our teeth and holding them in place. Since our permanent teeth are usually in place before the end of grade school, the long-term health of the gingivae is crucial for retaining our teeth for the duration of our adult lives.

Gingivitis is a condition marked by swelling of the gums. The swollen gums become red and sensitive, and bleed freely when brushed. Gingivitis may be the result of pregnancy or diabetes, but the most common cause is poor oral hygiene. When the oral cavity is not cleaned regularly, food debris and bacteria begin to accumulate, especially along the gum line and between the teeth. This material is called *plaque*. Over time, the plaque hardens to form *calculus*, commonly called *tarter*. This hardened material is abrasive to the gums, causing them to redden and become inflamed. The tartar also houses bacteria, which as they digest the sugars and other food material between the teeth, produce toxins that further aggravate the gums. The aggravated gums bleed freely. The accumulation of plaque may also cause halitosis, or bad breath.

Gingivitis is easily treatable by returning to an aggressive oral hygiene regiment that includes regular brushing of the teeth and gums, and the use of dental floss or dental tape to remove material between the teeth. A dental hygienist can best remove the tartar using dental tools that remove the irritating agents from the gum line. The hygienist can also instruct the patient on proper oral care. Plaque and tartar can accumulate even in people who are highly dedicated to personal oral hygiene, so regular (twice per year) dental checkups and cleanings are highly recommended.

However, if left untreated, gingivitis can cause a potentially much more harmful disease called *periodontal disease*. Periodontal disease is the second stage of gum disease, and it marked when the gums begin to separate from the tooth and the infection of the gums begins to affect the tooth itself (see Figure 8.1). These pockets are prime breeding grounds for bacteria, but unlike gingivitis, these bacteria can't easily be removed from the area around the tooth. As the disease progresses the gums separate farther from the tooth and the periodontal ligament and alveolar bone may become infected. If left untreated, the tooth will loosen and tooth loss will result. A periodontist (a dentist who specializes in the treatment of periodontal disease) has a num-

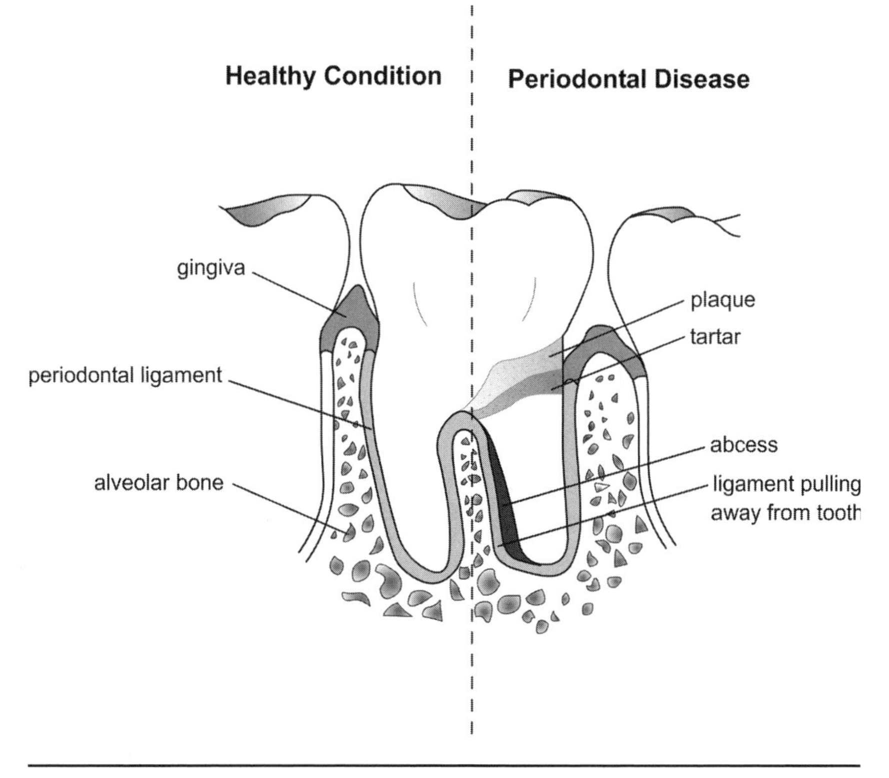

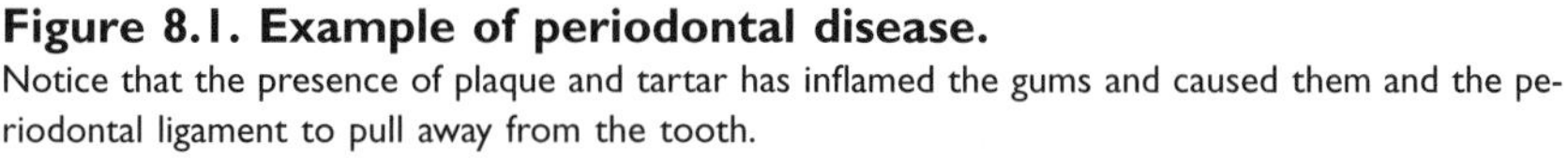

Figure 8.1. Example of periodontal disease.
Notice that the presence of plaque and tartar has inflamed the gums and caused them and the periodontal ligament to pull away from the tooth.

ber of treatment options for people suffering from periodontal disease based on the severity of the case. These include surgery to repair the gums and deep cleaning of the pockets.

Since one of the underlying causes of gingivitis and periodontal disease is the bacteria normally present in the mouth, some researchers have been investigating the use of a vaccine to increase personal protection against diseases of the gums. Because vaccines must assist the immune system to be effective in fighting a specific pathogen, researchers have focused on the bacteria *Porphyromonas gingivalis*, one of the major contributors to periodontal disease. Initial primate trials by scientists at the University of Washington have demonstrated the use of the vaccine may reduce the loss of bone tissue. At this time research is still underway on the development of a human form of the vaccine.

The extent of periodontal disease and gingivitis may be determined several ways. First, an x-ray of the oral cavity is done to examine the condition of the roots of the teeth and the surrounding alveolar bone. Second, a dentist or periodontist will use a metal probe to examine the space between the teeth and gums. Normally this space measures about 4⁄100–12⁄100 of an inch (1–3 millimeter), any more than this is an indication of gum disease.

While everyone is susceptible to gum disease, studies have shown that people who smoke have a much higher rate of periodontal disease than non-smokers. Other factors that increase the risk for periodontal disease are diabetes, pregnancy, anti-depressant medications, stress, vitamin-C deficiencies (such as scurvy), general malnutrition, and any disease that compromises the immune system. Recent studies have indicated that there may be a genetic component that makes a portion of the population susceptible to periodontal disease, although the gene or genes responsible have yet to be identified.

DENTAL CARIES

Dental caries, or cavities, are a common problem of the teeth. They are the result of acid formed by oral cavity bacteria as they break down sugars. The breakdown of carbohydrates and sugars in food by the bacteria may start as quickly as twenty minutes after eating. The acid produced by the bacteria breaks down the enamel coating on the teeth, producing a dental cavity. Most cavities do not cause discomfort in the individual until the inner areas of the tooth are penetrated and the pulp and nerves of the tooth are affected.

Currently, to remove a cavity from a tooth, a dentist uses a small high-speed drill to purge the tooth of the decayed enamel. Sometime in the next few years it is likely that the Food & Drug Administration (FDA) will approve the use of a **laser** for cavity removal, forever ridding the dentist's office of the sound of high-speed drills. However, the use of lasers is currently confined to work on soft tissues, such as the gums. Once the material is removed, a filling is fitted to the tooth.

Fillings may be manufactured from a number of different compounds, but the two most common are amalgam and resin fillings. Gold is also sometimes used, but its cost limits gold's widespread use. Amalgam fillings are a composite of silver and mercury, but may also contain copper, zinc, or tin. This type of filling has been used for over 150 years and is commonly called a metal filling. Due to the nature of the amalgam, it is easy to fit into the drilled cavity and forms one of the strongest bonds with the tooth. Recently there has developed a concern that the mercury found within the filling may contribute to diseases such as **multiple sclerosis, Alzheimer's disease**, and autism. However, the mercury in these fillings is exceptionally small and is bound as an alloy to the other metals. Furthermore, claims that the mercury within the fillings is responsible for disease is disputed by the American Dental Association (ADA) and to date no links to any disease have been found by the FDA or the Centers for Disease Control (CDC). People who are concerned about the mercury content of their fillings should consult with their dentist for more information.

A second, newer form of filling is called the composite resin dental filling. This form of filling uses a plastic compound, or resin, to fit into the cavity. It functions in much the same way as a traditional filling, but unlike the amalgams, the resin fillings look more like a natural tooth. Ceramic fillings may also be used in some cases.

In addition to fillings, dental sealants may be used to coat the teeth and protect them from decay. The sealants, which are usually plastic coatings, are applied to the surface of the molars. A single application usually lasts several years.

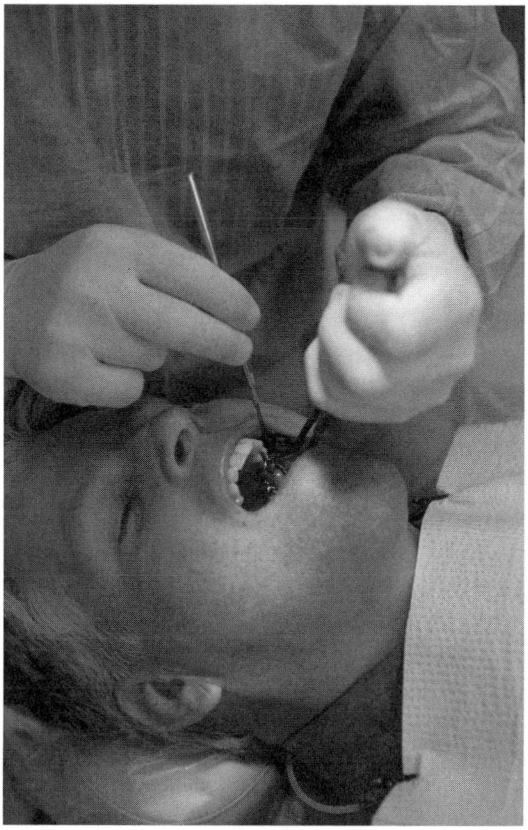

A visit to the dentist. © Photodisc.

ROOT CANALS

Root canals, or endodontics as the procedure is called by dental professionals, represents a more invasive treatment of decay within a tooth. While dental fillings repair the enamel of the tooth, the purpose of endodontic treatment is to remove infected material from beneath the enamel of the tooth. This is commonly infected pulp or nerve tissue (see Figure 2.4), although the process can also be performed if an abscess, or infected cavity, is located beneath the tooth.

Dentists frequently refer patients who require endodontics to a dental specialist called an *endodontist.* Before a root canal begins the entire nerve is numbed using an anesthetic. Following this, a hole is made in the crown of the tooth, and the infected pulp and nerve material is removed. The entire tooth is cleaned and the cavity is filled using a rubber-like compound. A filling is then placed over the crown.

On occasion, the damage to the tooth is too deep to be reached through the crown of the tooth. The dentist or endodontist may then elect to perform a treatment called an *apicoectomy.* In this procedure, the gum below the tooth is cut and the infected tissue beneath the tooth is treated. The need for endodontics, fillings, and periodontal treatment are usually the result of an ineffective oral hygiene regime. Proper care of the teeth and gums is essential for maintaining the health of the teeth and gums.

DISEASES AND AILMENTS OF THE ESOPHAGUS

The esophagus is essentially a conduit between the oral cavity and stomach (see Chapter 2), and many of the ailments of the esophagus are due to underlying problems with the physiology of the stomach. (Chapter 9 will examine these diseases in greater detail.) However, the esophagus is not without its own ailments. Most ailments of the esophagus, with the exception of esophageal cancer (see Chapter 12), are the result of changes in the motility of the bolus as it moves down the esophagus, or a problem with one of the valves or sphincters that isolate the esophagus from the stomach.

Heartburn

One of the more common problems of the esophagus, affecting up to 7 percent of the U.S. population, is heartburn. One of the primary causes of heartburn is gastroesophageal reflux disease (GERD). Heartburn is characterized by a burning sensation that begins under the sternum and typically moves upward into the neck or throat region. While its symptoms are frequently confused with the symptoms of a heart attack, GERD has no relationship to the heart with the exception of the location of the discomfort.

GERD occurs when the gastroesophageal sphincter (see Chapter 2), sometimes called the cardiac sphincter, does not completely isolate the esophagus from the stomach. Under normal conditions, this valve allows food and liquids to pass into the stomach, but then closes to prohibit the acidic gastric juice of the stomach from coming in contact with the delicate tissues of the esophagus.

Under certain conditions, the gastroesophageal sphincter weakens. This may be the result of the natural aging process, a hiatal hernia of the stomach (see Chapter 9), or certain foods that irritate the sphincter. Examples of foods that may create problems are fried/fatty foods, coffee, alcohol, and chocolate. The types of food that produce the problem may vary by individual. In addition, some people experience discomfort when in a horizontal position, especially when the head is at the same level as the rest of the body.

While heartburn is not a serious ailment, it left untreated it can permanently damage the lining of the esophagus. In some cases long term exposure on the part of the esophageal tissues to acids is believed to cause a form of esophageal cancer (see Chapter 12), although this is rare. Treatments usually include avoidance of problem foods, a decrease in alcohol and coffee consumption, changes in the quantity of food that is consumed at a single time (especially before bed) and the use of antacids to neutralize stomach acid. It should be noted that most physicians discourage the long-term use of antacids since this may create calcium and magnesium imbalances in the body.

If a physician suspects problems with the gastroesophageal sphincter, he or she may recommend a variety of tests, including x-rays. Recently physicians have begun using small cameras inserted directly into the gastrointestinal track to examine the structure of diseased organ. This procedure is called *endoscopy* and utilizes a small camera called an **endoscope**. Chemical tests may also be run to determine the amount of acid entering into the esophagus. If the condition is severe enough, or persists for an extended period of time, a doctor may prescribe a number of drugs that limit the stomach's ability to produce acid. Only in very severe cases is surgery an option.

Dysphagia

Dysphagia is an ailment whose symptom is a difficulty in swallowing. Although there are many conditions that may cause this problem, including decreased saliva output by the salivary glands (see Chapter 11), it usually indicates a problem with the esophageal muscles that are decreasing the motility of the bolus as it proceeds to the stomach. Dysphagia may be a sign of esophageal cancer (see Chapter 12) and therefore anyone with a history of this problem should consult with their physician.

In some cases GERD may damage the tissues of the esophagus to the extent that the movement of the bolus is slowed considerably. This may be due to an incorrect timing of the normal peristaltic contractions in the esophagus, which may delay the transit time of the bolus considerably. In other cases the lower esophageal sphincter does not relax completely, and thus food backs up into the esophagus. This condition is sometimes called *achalasia*.

Diagnosis of dysphagia is conducted in much the same manner as GERD examinations. Physicians may use an endoscope to examine the structure of the esophagus, or an instrument called a **manometer** to measure the pressure of the tissues. The doctor also may perform a procedure called an *esophagram*: the patient swallows a solution of **barium** while a **radiologist** examines the structure of the tissues in the esophagus and upper GI tract.

Treatment of dysphagia varies depending on the cause. If GERD is the culprit, the medications that reduce the production of acid by the stomach may be used. If achalasia is diagnosed, then a balloon is passed down into the esophagus and inflated to put pressure on the gastroesophageal sphincter.

Diseases and Ailments of the Stomach

Of all of the organs of the digestive system, the average person is most familiar with medical problems related to the stomach. Indigestion and nausea are household terms, and over-the-counter (OTC) medications to treat these ailments are a billion-dollar-a-year industry in the United States alone. However, ailments and diseases of the stomach can be much more numerous and significant than simple nausea. The stomach serves an important role in the preprocessing of the food so that it can be properly digested and absorbed by the small intestine. Furthermore, the hydrochloric acid acts as a physical barrier of the immune system, thus protecting the more delicate intestines from parasites and bacterial infections. Medical conditions that limit the effectiveness of the stomach in these roles can have severe consequences on the overall health of the gastrointestinal system. (One potentially fatal disease of the stomach, cancer, will be covered in more detail in Chapter 12.)

NAUSEA AND VOMITING

Of all of the ailments that can affect the stomach, the causes of nausea and vomiting can be the most difficult to pinpoint. Nausea is typically described as a general feeling of discomfort within the gastrointestinal system, usually centered on the stomach. Anyone who has ever experienced seasickness understands the feeling of nausea. In fact, the word *nausea* is derived from the Greek word meaning "seasickness." Vomiting is the process by which the contents of the stomach and duodenum are forcibly expelled

X-ray of a normal abdominal region, including the esophagus, stomach, duodenum, and small bowel. © SIU BioMed/Custom Medical Stock Photo.

from the body. Vomiting is usually accompanied by a feeling of nausea, but nausea need not result in vomiting.

Nausea may be caused by a variety of factors, not all of which are related directly to the stomach. Some ailments of the small intestine, especially those involving decreased motility (see Chapter 10) can induce a feeling of nausea. Nausea may also be caused by the nervous system. A common example is motion sickness, such as that experienced in a moving vehicle such as a car, boat, or plane. During motion sickness, the central nervous system (CNS) receives conflicting information from various sensors in the body. For example, if you are flying in an airplane some receptors signal the brain that you are sitting still in the cabin, while the **vestibular apparatus** within the ear detects acceleration or changes in motion. The result is frequently nausea, although the interaction between the nervous system and stomach is still not understood. Astronauts frequently experience nausea in space as a result of weightlessness. Space agencies such as the National Aeronautics and Space Administration (NASA) are actively investigating the causes of motion-related nausea to develop cures for long-term space flight.

Nausea may also be caused by diet. Overeating can create temporary feelings of nausea; since the stomach may only contain a maximum of 2.11 quarts (2 liters) of food (see Chapter 2), any overloading of the stomach may create an unsettled feeling. Alcohol is an irritant to the stomach lining and may also cause feelings of nausea, especially if the alcohol is consumed on an empty stomach. As a toxic substance, alcohol can also induce vomiting.

Other physiological factors that may cause nausea are stress, sensory input from undesirable sources (sight of blood, smell), migraine headaches, and hormonal changes brought on by pregnancy. In addition, medical procedures such as general anesthesia or chemotherapy frequently produce nausea in the patient, as do many prescription drugs. Most pharmaceutical companies perform a significant amount of research on whether their drugs will create nausea (since this may reduce the patient's desire to take the drug) and make recommendations on how to reduce the severity of nausea while on the medication.

Vomiting is a physiological response that is typically associated with nausea. In order for vomiting to occur, the pyloric and gastroesophageal sphincters (see Chapter 3) must relax. Frequently the movement of chyme in the duodenum is reversed so that intestinal contents are returned to the stomach. The stomach itself does not contract; rather the diaphragm and abdominal muscles tighten to place pressure on the stomach, propelling the contents upward. If the material is coming from the duodenum, it will frequently have a yellow tint since it has been mixed with bile salts. Vomiting that contains intestinal material is not as rare as people think.

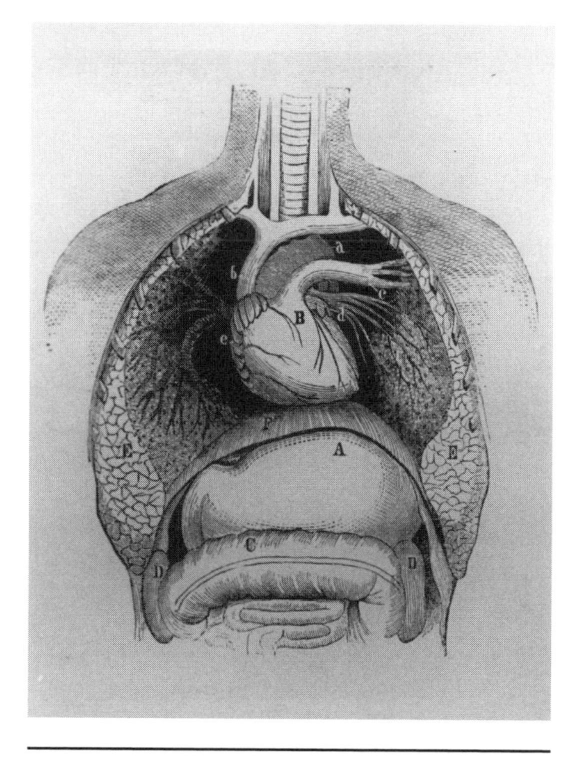

A distended stomach and colon. Wood engraving from S. S. Filch, *A popular treatise on diseases of the heart*, New York 1859. © National Library of Medicine.

Vomiting is the body's attempt to remove toxins from the stomach and upper regions of the small intestine. However, it may be caused by other factors, including intense nausea, trauma to the body, or an emotional response. Historically, until the twentieth century, vomiting was viewed as a beneficial action that could be utilized to rid the body of toxins. While induced vomiting is still used in some cases of accidental poisoning (but only when directed by a physician), most medical professionals do not share the view of their historical counterparts. Vomiting can lead to damage to the esophagus, dehydration, and nutritional problems, especially if it persists for an extended period of time. Bulimia, a psychological disorder in which people self-induce vomiting following a meal, can cause severe degradation of the esophagus and oral cavity.

Nausea and vomiting are usually accompanied by other physiological factors, including increased heart rate, increased saliva production, constriction of blood vessels in the skin, and the desire to defecate. In severe cases, a doctor may prescribe medications to reduce vomiting, but in most cases, patients should remain calm and drink plenty of water to replace lost fluids.

PHYSICAL PROBLEMS OF THE STOMACH

Peptic Ulcers

One of the more common physical problems of the stomach is an *ulcer*. An ulcer is an open sore in the lining of the digestive track. The mucosa layer of the organ is removed in the area of the sore, and the lower tissue layers (submucosa, muscalaris) are exposed to the lumen. Ulcers may occur in either the stomach or intestines. Ulcers in the stomach are called peptic ulcers, and, although similar in structure, they may arise from different factors than intestinal ulcers. Intestinal ulcers, usually called duodenal ulcers, are covered in additional detail in the Chapter 10.

The mucosa layer of the stomach secretes a thick mucus layer to protect the delicate tissues of the stomach from the effects of the hydrochloric acid and strong proteases present in the lumen. When this layer is degraded or removed, the acid and proteases of the stomach begin to digest the exposed tissue. The result is a painful burning sensation that usually begins within three hours following a meal. Eating usually causes the pain to subside.

Until the 1980s, peptic-ulcer disease was believed to be caused by irritations to the mucosa lining by alcohol, medications, stress, spicy foods, or smoking. However, in 1982, a bacteria, then called *Campylobacter pylori*, was isolated (see following section regarding parasites). This bacteria, now called *Helicobacter pylori*, is probably associated with over 80 percent of all peptic ulcers. Physicians no longer accept that stress and spicy foods cause peptic ulcers. For those cases not associated with the bacteria, attention has focused on smoking, non-steroidal anti-inflammatory drugs (NSAIDs) such as aspirin and ibuprofen, and a general excess of stomach acid. A rare pancreatic disorder called Zollinger-Ellison syndrome (see Chapter 11) may incorrectly signal the stomach to release hydrochloric acid, causing ulcers.

Treatment for a non-bacterial peptic ulcer may take several forms, most of which involve assisting the body's natural healing process. Antacids are frequently used to reduce the immediate symptoms of a peptic ulcer, but in general do not aid the body in repairing the problem. A new class of drugs called the proton pump inhibitors (an example is the brand-name Prevacid), reduce the production of hydrochloric acid by the stomach. These drugs work over time and do not provide immediate relief. Another drug therapy involves the use of H_2 blockers, which also reduce stomach acid production. These drugs inhibit **histamine,** thus decreasing the ability of the cells to secrete compounds such as acid.

Ulcers should be taken seriously, and a person should consult a physician if that person suspects he or she has an ulcer. Ulcers may block the pyloric sphincter and are being associated with forms of stomach cancer and lymphoma (see Chapter 12). Ulcers are usually accompanied by gastritis (see following section on gastritis). While many ulcers may heal themselves over time, any sign of bleeding is a reason for immediate medical attention.

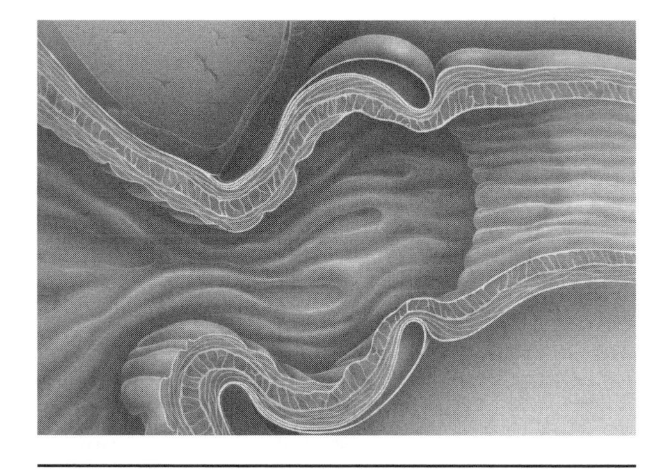

Illustration of a hiatal hernia. © K. Sommerville/Custom Medical Stock Photo.

Hiatal Hernia

Another problem that affects some people is a *hiatal hernia*. A hernia is when a section of the abdominal wall weakens or tears, allowing tissue from another organ to protrude into the space. The majority of hernias involve the intestines (discussed in more detail in Chapter 10). Normally, hernias are not life-threatening, unless the blood supply to the tissue is cut off. This is called a *strangulated hernia* and can produce a severe systemic infection. In the case of a hiatal hernia, a small tear or bulge in the diaphragm (located above the stomach) allows a small portion of the stomach to protrude into it. This usually occurs where the esophagus penetrates the diaphragm, close to the gastroesophageal sphincter. When this happens acid from the stomach becomes trapped between the gastroesophageal sphincter and the herniated section of stomach. This produces many of the same symptoms as GERD (see Chapter 8). Hiatal hernias may be caused by sudden strain on the stomach from lifting, severe coughing, or vomiting. As people age, the tendency for a hiatal hernia to develop increases, although people of any age are susceptible. In most cases of hiatal hernia, no treatment is necessary, unless the pain is severe or the associated GERD is damaging the esophagus. Antacids can relieve the symptoms of a hiatal hernia, but cannot correct the underlying physical problem.

Gastritis

Gastritis is the inflammation of the mucosa layer of the stomach. As is the case with any inflamed tissue, gastritis may produce bleeding and pain. Pain associated with the upper GI tract is commonly called *dyspepsia*. Gastritis may be caused by a number of factors and is usually present in the tissues surrounding ulcers. Bacterial infections, specifically those caused by *H. pylori* (see next section), frequently result in gastritis. In these cases, the use of an antibiotic to control the bacteria will allow the tissue to heal naturally. The use of non-steroidal anti-inflammatory drugs (NSAIDs) such as aspirin and ibuprofen, may cause gastritis as well. In other cases, an excess of hydrochloric acid wears away the protective mucus layer, allowing the gastric juice to come in direct contact with the mucosa. This irritation causes gastritis. In this case medications that reduce hydrochloric acid production, such as the proton pump inhibitors and H_2 blockers, reduce the irritation on the tissue. The most common causes of gastritis are smoking and alcohol use. While the mechanism by which smoking causes gastritis is not completely understood, the toxins in the cigarette smoke are believed to weaken the stomach lining, predisposing it to irritation. Alcohol has the same effect.

In a rare number of cases, gastritis is the result of an autoimmune response by the body against the cells of the mucosa. These cases are frequently differentiated from others by the fact that the damaged tissues have poor absorption properties. The discovery of vitamin B_{12} in the early twentieth century began as the study of pernicious anemia, which is now known to be an autoimmune disease. Patients with pernicious anemia are unable to produce the intrinsic factor for vitamin B_{12} absorption, resulting in anemic conditions (see Chapter 7).

Gastritis has been linked to both stomach cancer (see Chapter 12) and a rare form of lymphoma that originates in the lymphatic cells surrounding the stomach. The mechanisms by which both of these originate is still under investigation.

PARASITES

At first glance, the environment of the stomach, with its pH of around 2.0 and presence of strong protease enzymes, would make it an unlikely place for a parasitic infection. However, the truth is that some bacteria have evolved elaborate mechanisms of living in the stomach (see "Living at a pH of 2.0"). Still others have developed physiological adaptations that allow them to transit the stomach environment and take up residence in the intestines (see Chapter 10).

For the longest time stomach ulcers were considered to be simply a physiological response by the body due to diet, lifestyle or irritants such as alcohol. While these factors still play an important role in the development of many types of ulcers, since the early 1980s attention has turned to a species of bacteria called *Helicobacter pylori.*

In 1981, two researchers at the Royal Perth Hospital in Australia, J. Robbin Warren and Barry Marshall, isolated a bacteria from patients with gastritis and peptic ulcers. Initially called *Campylobacter pyloris,* the name was changed to *Helicobacter pylori* once it was realized that the species of bacteria they were examining had never previously been identified. This spiral-shaped, curved bacteria had the ability to survive in the conditions of extreme pH (2.0) and low oxygen concentration of the stomach.

Despite the fact that *Helicobacter pylori,* or *H. pylori* as it is usually called, was present in patients with gastritis, the medical community was slow to

Living at a pH of 2.0

For environmental conditions, a pH of 2.0 for a bacteria would be like us taking a shower using battery acid. How then does *Helicobacter pylori* survive at conditions that are lethal to our own cells? First of all, *H. pylori* is not unique; to date scientists have identified other species of the genus in the intestinal tracts of other organisms. Second, it appears that *H. pylori* does not love acid. Rather, it creates a microenvironment around itself through the action of an enzyme called *urease*. Urease breaks down urea (which is naturally found in all body fluids) into carbon dioxide and ammonia. It is believed that the ammonia is then utilized to neutralize the hydrochloric acid in the immediate vicinity of the cell. Finally, *H. pylori* thrives in the low-oxygen content (approximately 5 percent) of the stomach, which explains the early problems with culturing the bacteria in the lab. In fact, normal oxygen atmospheric concentrations of 21 percent are detrimental to *H. pylori*'s growth. Many scientists believe that if *H. pylori* has evolved the mechanisms to live in this inhospitable environment, then there may be other yet unidentified bacteria in the human body that create other inflammatory diseases.

accept that the bacteria was causing the disease. Many believed that the bacteria was a secondary infection, and was simply using the damaged tissue to avoid the hydrochloric acid of the stomach. Additional animal studies indicated that not only did *H. pylori* cause ulcers and gastritis in animal models, but also that the use of antibiotics would cure the infection and the stomach ailments. The medical community remained skeptical until in a highly unconventional experiment Marshall drank a sample of *H. pylori* and gave himself gastritis. He then treated the ailment with antibiotics.

Almost 80 percent of the cases of gastritis and peptic ulcers are the result of *H. pylori* infections. Almost half of the world's population carries *H. pylori*. However, for reasons not completely understood, not everyone infected with *H. pylori* gets gastritis or ulcers. *H. pylori* is believed to be transmitted by contaminated water or close contact, and thus is prevalent in developing countries around the world.

Since *H. pylori* is a bacterial infection, the body produces antibodies against the bacteria. This occurs in anyone who carries the bacteria, regardless of whether they display the symptoms. However, the antibodies are not able to control the infection, probably due to the location of the bacteria within the lumen of the stomach. Since antibodies are present, it is possible using a blood test to determine if a patient's gastritis is the result of a bacterial infection or another factor. Using this information, a physician can establish a treatment regimen targeting either the bacteria using antibiotics, or the production of acid using medication.

Infections of *H. pylori* have been linked to an increased rate of stomach cancer, as well as lymphomas in the lymphoid cells of the stomach tissues. *H. pylori* can also cause ulcers in the small intestine (see Chapter 10). Due to the ease of treatment, patients with recurring gastritis are usually encouraged to consult with a physician as to whether they are infected with *H. pylori*.

Diseases and Ailments of the Small and Large Intestine

In Chapter 3, we defined the lower GI tract as being all digestive organs below the level of the pyloric sphincter. This includes the small and large intestine. The liver, pancreas, and gall bladder all act as accessory organs to the lower gastrointestinal tract (Chapter 4). Ailments and diseases specific to those accessory organs will be covered in more detail in Chapter 11. Since cancer is a problem of both the small and large intestine (as well as the accessory organs), it will be discussed in its entirety in Chapter 12.

DISEASES AND AILMENTS COMMON TO THE SMALL AND LARGE INTESTINE

Parasites of the Intestinal Tract

As mentioned previously, the lumen of the gastrointestinal tract represents the location of the extracellular processing of nutrients. This means that the lumen is actually outside the tissues of the body, even though it is contained within the physical boundaries of the human body. This is an ideal environment to support life, since it supplies a constant temperature and a readily available supply of nutrients. It is no surprise then that there are a number of organisms, mostly bacteria, which naturally live within this area. Many of these cause no medical problems for humans. These are called the natural *flora*, and their function in nutrient processing has been already discussed (see Chapter 3). This section will examine the more common intestinal parasites, organisms that cause medical problems in humans. It is not meant to be a complete list of pathogenic organisms, but rather give an indication of what, if given the opportunity, can infect the intestinal tract.

MICROBES

Microscopic organisms include bacteria and larger single-celled organisms called the **protistans**, sometimes called the **protozoans**. Despite the harsh environment of the stomach, with its strong protease enzymes and high acidity, some microbes still manage to get into the small and large intestines. There they encounter the next line of defense: a well established natural flora of bacteria which inhibit the growth of the pathogenic organisms. For many microbes that enter into the digestive tract, this is the end of the line, and the individual is never aware that an infection has been stopped before it began. However, once in a while an organism can evade the defenses and establish itself in the gastrointestinal system.

CHOLERA

The disease cholera is caused by the bacteria *Vibrio cholerae*. The symptoms of cholera include diarrhea, vomiting, and abdominal cramping. As is the case with the majority of diarrhea-producing intestinal ailments, dehydration is the major concern. A cholera infection begins when a person ingests live *Vibrio cholerae* bacteria from fecal-contaminated food or water. The bacteria then bind to the cells of the small intestine, releasing a toxin. It is this toxin that causes the release of fluids by the cells, and the resulting symptoms of diarrhea. Diarrhea caused by cholera has a very high water content.

Cholera is not a life-threatening disease if treated. Treatment includes the use of antibiotics to assist the body's immune system, as well as an increase in water and electrolyte consumption to replace losses due to diarrhea. Without treatment, the disease can be fatal. While there have been no major outbreaks of cholera in the United States since the early years of the twentieth century, the bacteria is still present in some water supplies and there have been isolated cases around the country. The disease is primarily a problem of developing countries, especially those that lack sanitation systems to purify drinking water.

Typical *Vibrio cholera*–contaminated water supply. © Centers for Disease Control and Prevention.

ESCHERICHIA COLI

Escherichia coli, or *E. coli* as it is usually called, in some ways is one of the most beneficial bacteria in the world. *E. coli* is a model organism for the study of bacterial genetics and physiology. Virologists often use *E. coli* as a host organism for their research. Furthermore, it is frequently used in biotechnology research due to the relative ease with which the organism can be cultured in the lab. From a medical perspective, *E. coli* represents an important component of the natural flora of the large intestine (see Chapter 3), where it is responsible for establishing a healthy environment in the colon, as well as assisting in the production of some vitamins and sugars.

However, when one hears of *E. coli* in the news, it is usually in reference to an outbreak of intestinal illness associated with contaminated food or water. In most cases these outbreaks are due to a strain called *E. coli* O157:H7. Unlike most forms of *E. coli*, O157:H7 produces a toxin in the intestinal tract that causes severe abdominal pains and diarrhea. At times the diarrhea may be bloody. However, some people are **asymptomatic** and can carry the pathogenic strain without showing symptoms.

E. coli O157:H7 infections are typically associated with the consumption of undercooked beef. The bacteria is naturally found in cattle, and in very rare cases the pathogenic strain may be present. Thorough cooking of the meat prior to consumption eliminates the bacteria. Since the bacteria is normally present in cattle, milk products may also be contaminated by *E. coli* O157:H7, but these are effectively removed by **pasteurization**. Another source of an *E. coli* infection is the ingestion of water that is contaminated with animal or human waste, especially if that waste contains O157:H7. Several incidents of this form of infection have occurred at water parks, swimming pools, and when contaminated water was used to irrigate crops.

For most people infection with *E. coli* O157:H7 is not fatal, although the disease can be dangerous if a significant amount of fluid or electrolytes are lost. The use of antibiotics can lessen the duration of the disease. However, in young children, primarily those under five years of age, an infection with *E. coli* O157:H7 can become systemic. In this case the most common result is a disease called *hemolytic uremic syndrome*. In hemolytic uremic syndrome, red blood cells are lysed, or destroyed, and the kidneys shut down. Around 5 percent of these cases are fatal. Precautions that will help avoid an *E. coli* infection include the following:

Thoroughly cook (to 160°F) all beef products prior to consumption.

Clean all surfaces and utensils that come in contact with raw meat.

Wash hands thoroughly after using the bathroom.

Isolate children with diarrhea from other children and family members.

Wash all fruits and vegetables.

Do not drink from unknown water sources.

Avoid swallowing water from swimming pools.

SALMONELLOSIS

Salmonellosis is an ailment of the intestinal tract caused by members of the *Salmonella* genus of bacteria. In many regards, an infection with *Salmonella* bacteria is very similar to an infection with a pathogenic strain of *E. coli.* Salmonellosis is characterized by abdominal pain and diarrhea. The bacteria is naturally carried by a number of animals, including birds. Most people who contract salmonellosis come in contact with the bacteria by either eating undercooked poultry products or improperly cleaning utensils that come in contact with fluids from poultry. Some cases have been reported from individuals who have ingested contaminated water supplies.

As is the case with *E. coli,* salmonellosis is treatable using antibiotics coupled to fluid and electrolyte replacement. In healthy individuals, the infection may run its course in a few days, without a need for antibiotics. Many people may not recognize that they have had a *Salmonella* infection, but rather may think that they have had an intestinal "bug." The disease can be fatal, but mostly only in individuals with compromised immune systems, or in the very young or elderly.

Prevention of salmonellosis involves reduction of exposure to uncooked poultry products or contaminated water. As was the case with *E. coli,* only thoroughly cooked poultry products should be eaten. Many cases of salmonellosis occur from the improper handling of cooking utensils. To prevent infection, all cooking utensils that come in contact with poultry preparation should be thoroughly cleaned and sanitized using hot water or bleach solutions. Never eat poultry using a utensil that was involved in the preparation of the raw meat.

WHIPPLE'S DISEASE

Whipple's disease is caused by the bacteria *Tropheryma whippelii.* The bacteria causes lesions to grow on the lumen surface of the intestines, most commonly the small intestine. This destroys the villi and microvilli and significantly reduces the absorption capabilities of the intestines. The symptoms of a *T. whippelii* infection include signs of malnutrition, such as weight loss and fatigue. Included with these may be diarrhea, intestinal bleeding, and pain in the abdominal region.

An infection of *T. whippelii* can also influence a wide variety of other body systems, including the immune system, circulatory system, and nervous system. Once diagnosed it is easily treated with antibiotics. However, the damage to the intestinal mucosa can take several years to repair, and frequently dietary supplements are necessary to ensure proper nutritional health during this time.

AMOEBAS

There are a number of amoeba species that may invade the gastrointestinal tract. Amoebas are able to gain access to the intestines by forming cysts, a dormant stage in which the amoeba frequently forms a tough, resistant covering. The most significant of these is the species *Entamoeba histolytica*. This organism exists in two major forms, a pathogenic strain and a nonpathogenic strain. The pathogenic version of the organism is responsible for the disease caused *amebic dysentery*, a worldwide disease that infects over 500 million people annually. An *E. histolytica* infection is typically acquired through the ingestion of contaminated food or water. Once the cyst passes through the pyloric sphincter into the intestinal tract, the amoeba enters the active phase of its life cycle. *E. histolytica* is a parasite of the lower small intestine (primarily the ileum) and the colon. There it forms ulcers on the mucosa layer. As it feeds on the live tissues, the immune system begins a response against the amoeba. However, *E. histolytica* possesses a defensive mechanism that destroys white blood cells, thus protecting it from the immune response.

Symptoms of amebic dysentery most often include bloody diarrhea, although abdominal pain and fever may also be present. Dehydration and malnutrition are major concerns of patients with amebic infections. Since the symptoms of an amebic infection may closely resemble those of colitis or other inflammatory diseases, it is frequently difficult to initially determine that an amebic infection is underway. Often a biopsy, or tissue sample, is required to confirm the presence of the pathogen. If left untreated, the ulcers may penetrate the intestines, resulting in a severe infection of the abdominal cavity. Furthermore, the amoeba itself may leave the intestinal tract and take up residence in other tissues of the body. Frequently, the secondary target of the amoeba is the liver, resulting in a disease called *hepatic amebiasis*. An amebic infection of the liver results in lesions, masses, or abscesses on the liver. In some cases, *E. histolytica* may infect the brain and lungs, causing abscesses in these tissues as well.

Amoebas in the intestines frequently form cysts which are passed out of the body along with the fecal material. It is these cysts that act as the agent of infection. While other organisms possess amoebas in their intestinal tract, human infections are the result of contamination of water or food supplies with human fecal material. The cysts of *E. histolytica* are often resistant to chlorination, a common mechanism of purifying water supplies.

Treatment for an amebic infection first involves identification of the organism involved, followed by the use of anti-amoeba chemicals, such as iodoquinol or metronidazole. The patient must usually also be treated for dehydration and malnutrition associated with the infection.

CRYPTOSPORIDIUM

Another parasite of the intestinal tract are species belonging to the genus *Cryptosporidium*. *Cryptosporidium* species belong to a group of single-celled organisms called the Sporozoans. As the name implies, these organisms form a **spore** at some point in their life cycle. In the case of *Cryptosporidium*, it is this spore that enables it to evade the hydrochloric acid of the stomach and establish itself in the intestines. One of the more common species that infect humans is *Cryptosporidium parvum*.

Cryptosporidium is another parasite that is transmitted by fecal-contaminated water supplies. Once the spore enters into the intestinal tract, it attacks the epithelial cells of the mucosal layer. The parasite actually enters into the tissue and becomes completely enveloped by it. The sporozoan then grows and releases additional organisms into the lumen of the intestines. The presence of the organism in the mucosal tissue causes the tissue to release large amounts of water and electrolytes. It is not clear what mechanism enables the organism to cause this reaction, and to date no toxins have been identified. The result is diarrhea and often dehydration.

In healthy individuals the infection often lasts only a few days, although it can persist for up to one month. With the exception of possible dehydration, there are few other side effects of the infection. In fact, many people may not realize that they have been infected with *Cryptosporidium*. The problem lies with young children, the elderly, or those with compromised immune systems. In these individuals the infection is frequently more severe, and the complications from prolonged diarrhea more pronounced.

Cryptosporidium is a major concern of major municipal water treatment facilities, since the spores of this organism are difficult to remove by filtration and almost always resistant to chlorine treatment. There have been several major incidents of *Cryptosporidium* infections in cities, the most famous of which is probably the incident that occurred in Milwaukee in April 1993. An estimated 403,000 individuals were infected with *Cryptosporidium* in this one outbreak. Typically these large outbreaks are the result of floods that overrun waste-water treatment plants or the release of untreated sewage into public waterways. *Cryptosporidium* can also be a problem in rural areas, where private wells may be infected by aging or inadequate septic systems.

GIARDIASIS

The intestinal disease giardiasis is caused by a microbe named *Giardia lamblia* (sometimes called *Giardia intestinalis* or simply *Giardia*). *Giardia* is another intestinal parasite that occurs due to the ingestion of fecal contaminated water supplies. Like the others previously mentioned, *Giardia* causes severe diarrhea, resulting in the loss of both electrolytes and fluids.

Giardia is an interesting organism in that it was first identified by the pioneer microscopist Anton van Leeuwenhoek (1632–1723) in the seventeenth century using a microscope of his own design. This makes *Giardia* the first microscopic parasite to be identified, although its role in disease would not become apparent for several centuries. It is also remarkable in its ability to resist both the harsh environment of the stomach and artificial mechanisms of water purification such as filtration and chlorination. The organism avoids damage by the formation of a tough, protective outer covering called a cyst and the use of a variety of detoxification enzymes. The cyst is highly resistant to severe chemical environments and since the cyst only averages around 12 micrometers in length, it is exceptionally difficult to filter. Unless a filter is rated 1 micron or less, it is unlikely to removes the cysts.

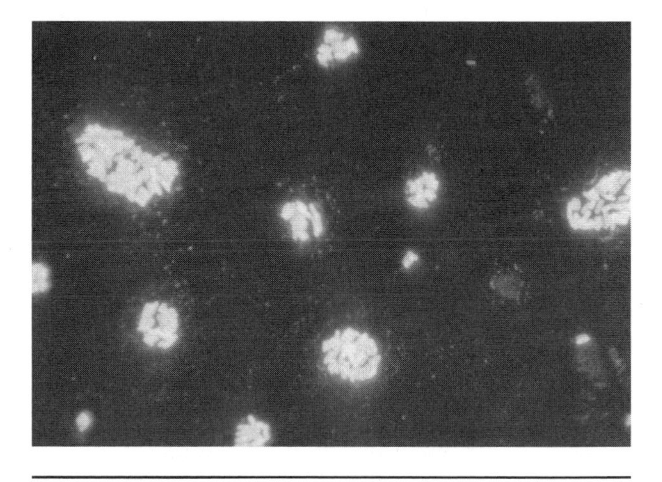

A photomicrograph of a positive indirect immunofluorescence test checking for the presence of *Giardia lambii*. © Centers for Disease Control and Prevention.

It is estimated that *Giardia* is responsible for close to 300 million cases of intestinal diarrhea per year, making it one of the most potent intestinal parasites of humans. One of the major sources of infection is the ingestion of recreational water, such as that found in swimming pools or public freshwater beaches. A developing cause of *Giardia* infections is the contamination of private wells by inadequate or failing septic systems. In addition, it is possible to contract an infection by coming in contact with the fecal material of an infected person, such as may occur when changing a diaper. Since individuals are able to spread the disease for up to two weeks following an infection, and some people may be asymptomatic, it may not always be obvious which individuals are carriers of the organism.

The symptoms of a *Giardia* infection include diarrhea, abdominal cramping, and nausea. The symptoms take an average of one week to begin after coming into contact with the organism and, on average, the infection lasts between two and six weeks. The duration may be shortened by the use of medications. In general, a *Giardia* infection is not life threatening, except in the elderly and young children, as well as individuals with a compromised immune system. However, the dehydration and electrolyte loss associated with the diarrhea may create complications in any individual.

Individuals may avoid a *Giardia* infection by not drinking from recreational water supplies, washing hands thoroughly before preparing foods, and avoiding contact with both human and animal fecal material. Since family members may easily spread the disease, any persistent case of diarrhea in a member of the family should be brought to the awareness of a family physician to determine if *Giardia* is the culprit. Individuals who enjoy outdoor activities such as camping and hiking should be equipped with *Giardia*-eliminating filtration systems (not iodine tablets) if they plan to use local water sources.

PARASITIC WORMS

Parasitic worms actually may belong to two different groups of animals. Many parasitic worms belong to the group called the Nematoda, commonly called the roundworms. Others belong to the general group called the Platyhelminthe, which includes the tapeworms and their close relatives. This section provides examples of the more common parasitic worms.

Pinworm infections are caused by the nematode *Enterobius vermicularis.* Pinworms commonly infect the large intestine, although they may be present at other locations within the gastrointestinal tract. For most individuals, a pinworm infection produces no symptoms or indications that the individual is infected. However, one symptom that may occur is anal itching, which is best understood by examining the lifecycle of this parasite.

A pinworm infection begins with the ingestion of eggs from fecal contaminated material. This is referred to as the fecal-oral route. Once in the large intestine the eggs mature into adults and mate. The female pinworm then migrates to the anus, usually at night, to expel her eggs. The female may lay up to 10,000 eggs at the surface of the anus. The female dies following deposit of the eggs. Within 6 hours the eggs are able to infect a new host, and can persist for several weeks in the environment. The eggs present at the surface of the anus may then migrate back into the colon. Within four to six weeks the adults mature and the lifecycle is repeated.

Once diagnosed, pinworm infections are easily treated using medications such as mebendazole or pyrantel pamoate. A physician usually diagnoses a pinworm infection from a stool sample, although sometimes a form of adhesive tape is applied to the area around the anus to remove eggs. Pinworm infections are common in small children and decrease significantly with age.

Ascarsis lumbricoides (commonly called *Ascarsis*) is a roundworm belonging to the Nematoda class. Among all of the species of parasitic worms, *Ascarsis* is probably the most prevalent, infecting an estimated one-sixth of the world's population. In the United States it is estimated that as high as 20 percent of the childhood population may be infected. The signs of an *Ascarsis* infection include malnutrition and obstruction of the gastrointestinal tract, although it may be exceptionally difficult to diagnose in a healthy individual. Many people may currently have a roundworm infection and not be aware of it until they visit the doctor for an unrelated illness and undergo an examination of the gastrointestinal tract.

Ascarsis is sometimes called the large intestinal roundworm since it has been known to reach lengths in excess of 11.7 inches (30 centimeters). Females of this group are able to create 200,000 eggs in a single day. It is the presence of the eggs in the fecal material that provides the primary method of diagnosing a roundworm infection. It also serves as the mechanism by which the organism is transmitted between hosts. *Ascarsis*, like most roundworms, moves between hosts via the fecal-oral route, meaning that the fecal material from one individual is unintentionally ingested by another. This usually occurs from fecal contaminated soil, which can hold *Ascarsis* eggs for several months.

Like many of the intestinal worms, *Ascarsis* can have a complex life cycle. After being ingested, the eggs hatch in the small intestine where the organism develops into an adult and, if the infection is widespread, mates with other members of its species. As adults, the organism may grow to a size as to obstruct movement of the undigested material in the gastrointestinal tract. In some cases the organism may migrate to the liver, and then the lungs and heart. The organism then develops in the air sacs of the lungs (alveoli). They then migrate up the trachea and are swallowed into the stomach, thus eventually returning to the small intestine. While in the small intestine, the organism lives in the lumen against the mucosa layer. The worms do not physically penetrate the layer, and thus must continuously move upward to resist the peristaltic contractions of the small intestine. Once diagnosed, treatment for a roundworm infection involves the use of medications, such as mebendazole. Unfortunately, many medications only affect the adult worm, and thus multiple treatments are necessary to completely rid the host of infection.

Tapeworms (Cestoda) belong to a **phylum** of animals called the Playhelminthes. There are many forms of tapeworms (see Table 10.1), but all share a similar physiological structure and lifecycle. In most cases infection begins with the ingestion or uncooked meat or through an insect intermediate. Unlike the roundworms and pinworms, which were free-living in the lumen of the intestinal tract, tapeworms anchor themselves to the lining of

TABLE 10.1. Sources of Tapeworm Infections in Humans

Species Name	Source
Diphyllobothrium latum	Uncooked or undercooked fish products
Dipylidium caninum	Ingestion of fleas from infected cats/dogs
Hymenolepis species	Grains or cereals containing infected insects
Taenia saginata	Uncooked or undercooked beef products
Taenia solium	Uncooked or undercooked pork products

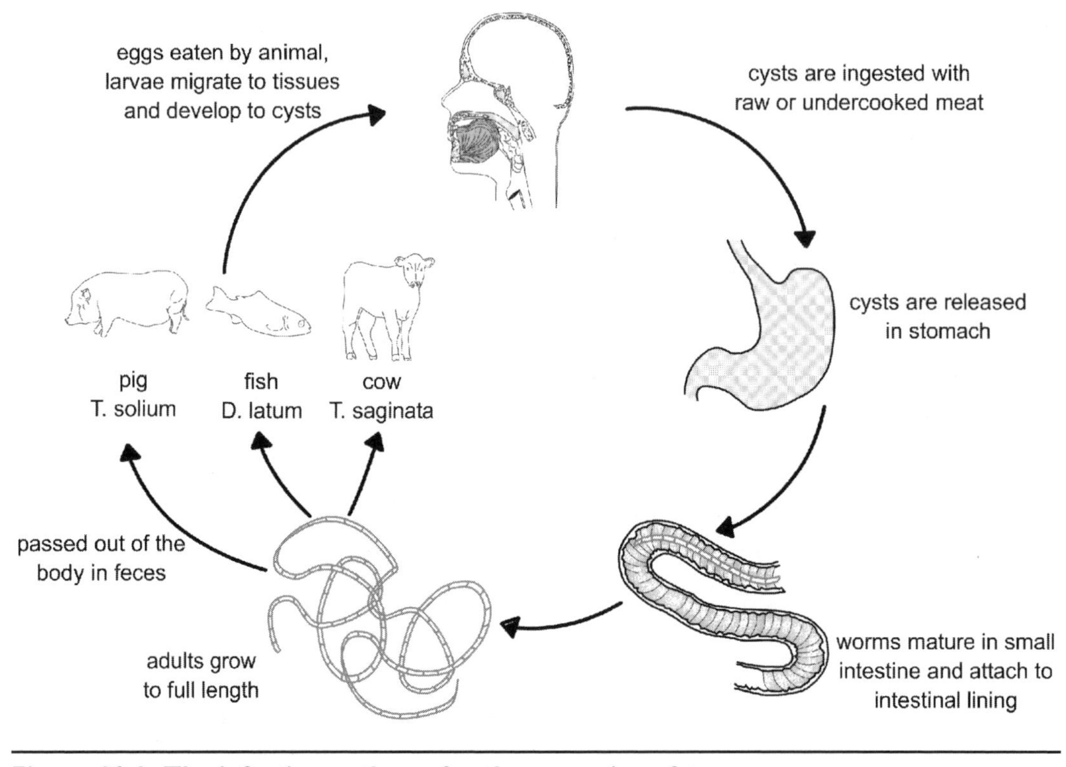

Figure 10.1. The infection pathway for three species of tapeworms.
Notice that all three species require a second host to complete their life cycle.

the intestines (usually the small intestine) using a set of suckers or hooks (depending on the species) called the *scolex*.

Infections usually involve a single tapeworm, not the thousands that may be possible with roundworm or pinworm infections. Because of this, tapeworms have an interesting form of reproduction (see Figure 10.1). The body of a tapeworm is divided into segments called *proglottids*. Each proglottid contains both male and female reproductive structures, and thus is **hermaphroditic**. Mating between two proglottids produces eggs, which are released into the environment with the fecal material. In most cases the eggs require an intermediate host (cattle, pigs, canines) to develop into larval form. In these animals the eggs form cysts, which gives the organism the ability to survive during their toxic passage through the stomach of the host. They then must be ingested by a host to complete development into adults. Adult tapeworms can be very long (over one meter) and have been recorded at lengths up to 9.1 yards (10 meters) (especially for *Diphyllobothrium latum*).

Tapeworm infections rarely cause problems for the host. In severe infections, there may be problems associated with malnutrition, usually of the B vitamins, but in most cases individuals are asymptomatic. Although they are longer, tapeworms are usually much less rigid than roundworms, and thus rarely cause an obstruction of the lumen. In fact, frequently the proglottids break off from the adult and are eliminated along with the fecal material. The presence of the proglottids in the stool is one diagnostic tool that may be used to indicate an infection. Once diagnosed, medications may be prescribed to dislodge the scolex from the intestinal wall. The adult is then expelled with the feces.

Physical Problems

CROHN'S DISEASE

There are several diseases of the intestinal tract that are characterized as being inflammatory diseases, meaning that inflammation of the tissues inhibits the ability of the organ to perform its role in digestion and absorption. Crohn's disease, sometimes also called *enteritis* or *ileitis*, belongs to this class of diseases. A similar condition resulting in inflammation of the tissues of the large intestine is called colitis. Crohn's disease differs from colitis in that it causes inflammation deeper into the tissues of the intestine.

Crohn's disease shares many of the same symptoms as irritable bowel syndrome (IBS) and colitis. Intestinal bleeding, diarrhea, and pain in the lower abdomen are all symptoms of both diseases. Crohn's disease may be accompanied by weight loss, fever, and rectal bleeding. It is frequently difficult to diagnose Crohn's disease since symptoms may appear to be cured and then reappear years later. Since intestinal bleeding may lead to anemia, a physician who suspects Crohn's disease may conduct blood tests to determine blood cell counts. The blood tests may also indicate an abnormally high number of white blood cells, a common occurrence in autoimmune responses. To determine the extent of the inflammation, the physician may conduct an upper gastrointestinal x-ray examination. However, the intestinal tissues in general do not resolve well using x-rays, so typically the patient will be asked to drink a special medication that contains barium. Barium is a metallic element that coats the interior surfaces of the gastrointestinal tract and allows for a greater resolution between diseased and normal tissues.

Crohn's disease is believed to be caused either by an immune response against some pathogen in the intestinal tract or an autoimmune response directed against the tissues of the intestine. The specific mechanism of this disease is still not well understood. However, recently researchers have identified a location on chromosome 16 that may hold a key to the problem of Crohn's disease. One of the genes at this location is involved in the immune response, and it is possible that people with defects in this gene may be susceptible to Crohn's disease, as well as other inflammatory intestinal diseases. If a defective gene that is responsible for at least some cases of the disease can be isolated, then that would explain the observations on the fact that forms of inflammatory diseases appear to be hereditary in some families. These cases may then be used to develop an effective treatment for the disease.

Besides causing discomfort in the patient, one of the more alarming effects of the disease is the problem of intestinal blockage. Unlike other inflammatory intestinal diseases, Crohn's disease typically affects deep into the tissues of the intestine, and not just the surface mucosal layer. This means that the inflammatory response is much more severe. In many cases the tissue becomes so inflamed that scar tissue forms and the lumen of the intestine becomes closed. Intestinal blockage can lead to rupture of the intestines, systemic infections, and even death. The widespread inflammation often extends past the intestines into other organs, including the skin, reproductive structures, and urinary system. Often this is in the form of small openings called *fistulas.*

Currently, treatment for Crohn's disease is complicated by the fact that the disease can disappear for months or years, only to reappear at the same level of severity as before. Medications are frequently used to treat the inflammation, but currently there is no cure for the disease. Treatments may involve the use of steroids to reduce inflammation or drugs containing a compound called mesalamine, which serves as an non-steroidal anti-inflammatory drug. Sometimes physicians will attempt to suppress the entire immune system, but this commonly leaves the patient open to infection from other sources.

New drugs such as methotrexate and cyclosporine are being used to make the immuno-suppression option more attractive. In addition, researchers are examining the chemical signals of the immune system to see if they can find a mechanism of halting the inflammatory response of the cells. One signal that shows promise may be interleukin-10. Additional research is focusing on the tumor necrosis factors (TNF), which appear to be involved in Crohn's disease. It is possible that medications may be developed that interfere with TNF in the body, and thus inhibit the inflammatory response. All of these treatments are still in the developmental stage.

Surgery is often presented as an option for severe cases of Crohn's disease. Inflammed sections of the small or large intestine may be removed, and the remaining sections reconnected. However, frequently the disease simply reappears in the connected sections. If intestinal bleeding or blockage is severe enough, a colostomy or ileostomy may be performed (see following section). Most physicians are reluctant to perform surgery unless there are complications in the disease, since removal of any portion of the intestine can create nutritional problems for the individual and is not guaranteed to prevent future occurrences of Crohn's disease.

COLOSTOMY AND ILEOSTOMY

In cases of severe problems with the small and large intestines, it may be necessary to either permanently bypass a section of the intestines or provide temporary rest for a section of the intestine; an operation to create the bypass is called an *astomy*. In these cases the intestine is diverted to the abdominal wall of the body where a small hole, or *stoma* is made to connect the intestine to the outside environment. Undigested material is then collected using a small bag or pouch on the outside of the body. If the diversion is at the terminal end (ileum) of the small intestine, it is called an *ileostomy*. If the diversion occurs along the length of the large intestine it is called a *colostomy*.

Colostomies and ileostomies may be necessary for a number of reasons, most common being diseases which render the large intestine unusable. In a temporary ostomy, the purpose is to give a section of the intestine a break from processing duties. This may be necessary in the case of severe diverticulitis, surgery involving the intestines, or colitis. Following recovery the procedure is then reversed and the gastrointestinal tract returns to its normal operation. If the portion of the intestine is permanently damaged, the surgery may be needed to permanently remove a section of the intestine, in which case the stoma serves to collect the undigested material.

Advances in procedures and equipment for ostomy patients practically ensures that there are no medical complications associated with this procedure, and patients can resume a normal, active lifestyle.

HERNIAS

Hernias are a common physical problem of the gastrointestinal tract, and can affect not only the intestines, but also the stomach and esophagus (see Chapter 9). A hernia is basically a weakening or tear of the abdominal wall that allows a portion of the gastrointestinal tract to extrude. This places pressure on the tissue, causing both pain and discomfort. While hernias can occur at any point along the intestinal tract, they most often occur in the lower portion of the abdomen. These are called *inguinal hernias* and are frequently caused by the lifting of heavy objects. In men, these hernias can protrude into the **scrotum**. In women they typically form a bulge in the lower abdomen or groin area. A second common form of hernia occurs in the area of the navel and is called an *umbilical* hernia.

Hernias frequently produce lumps on the surface of the abdomen. These lumps may have a burning sensation. They are not dangerous unless the blood supply to the tissue becomes cut off. This is a strangulated hernia and can quickly lead to medical complications such as infections. Fever or severe pain is a sign of a strangulated hernia.

Hernias do not have the ability to correct themselves and must be repaired surgically. Hernias may happen due to sudden abnormal movements of the abdomen, such as twisting and turning, and the lifting of heavy objects. The aging process increases the chances of a hernia, since the abdominal wall becomes naturally weaker over time.

AILMENTS AND DISEASES SPECIFIC TO THE SMALL INTESTINE

In Chapter 3, the small intestine was described as having three distinct regions. The upper third of the small intestine, connecting to the stomach, is the duodenum. This is followed by the jejunum and the ileum, which connects the small intestine to the large intestine by the ileocecal valve. Since the duodenum is in direct contact with the stomach, it shares many of the same ailments as that organ, including ulcers and inflammations. It is also important to note that the small intestine interacts directly with the gall bladder, pancreas, and liver, so problems with those organs frequently manifest themselves within the small intestine. (Diseases of these accessory organs will be covered in greater detail in Chapter 11.)

Duodenal Ulcers

One of the most prevalent ailments of the small intestine is an ulcer. Ulcers may occur in either the stomach (see Chapter 9) or small intestine. There are many similarities between gastric ulcers and intestinal ulcers (usually called duodenal ulcers). Due to the close proximity of the duodenum to the stomach, and the highly acidic content of the chyme as it passes through the pyloric sphincter, ulcers of the small intestine are usually confined to the upper reaches of the duodenum.

In almost all cases duodenal ulcers are caused by infections of *Helicobacter pylori* (*H. pylori*), a spiral shaped bacteria that was first discovered in the 1980s (see Chapter 9). *H. pylori* is a parasite of the stomach, and it is not yet clear as to why people with *H. pylori* in their stomach develop duodenal ulcers. However, once a patient has been diagnosed as having a *H. pylori* infection, treatment to remove the bacteria almost always results in the elimination of duodenal ulcers.

The symptoms of a duodenal ulcer are very similar to those of the gastric ulcers. A burning sensation in the upper abdomen and abdominal pain several hours after meals are common symptoms. Some patients experience nausea and vomiting as a result of the ulcer. If a duodenal ulcer is suspected a physician will typically check for *H. pylori* using a blood test, then examine the stomach and small intestine using either x-rays or an endoscope. Antibiotics can then be prescribed to eliminate the bacteria. Sometimes proton pump inhibitors and histamine-receptor antagonists (see Chapter 9) may also be used to reduce the amount of stomach acid entering into the small intestine.

Nutrition-Related Diseases

CELIAC DISEASE

Celiac disease, also called *celiac sprue* and *gluten-sensitive enteropathy*, is a condition in which the body produces an autoimmune response to a protein called *gluten*. Gluten is found in grains such as wheat, rye, barley, and oats. The presence of the food in the lumen of the small intestine causes the body to invoke an autoimmune response against the mucosal cells lining the villi. The result is the loss of the villi. Without villi, the absorption capability of the small intestine is drastically decreased, and patients frequently have problems with malnutrition.

Celiac disease is an inherited disorder, but a person who carries the trait usually does not display any symptoms until the disease is triggered by a stress to the body, such as a viral infection, pregnancy, surgery, or prolonged stress. The symptoms vary by individual, but frequently include chronic diarrhea, symptoms of malnutrition such as anemia, unexplained loss of weight, recurring problems with abdominal pain, flatulence and muscle cramps, just to name a few. If the symptoms vary or are inconsistent, the patient may be misdiagnosed until a pattern is detected in the symptoms. Since this is an autoimmune response, it is possible to screen individual for antibodies against gluten; if present, this confirms celiac disease.

Currently there is no cure for celiac disease, although researchers are actively investigating the possibility of drugs that inhibit the autoimmune response. Patients with celiac disease must adhere to a restricted diet that avoids grains such as wheat, rye, barley, and oats. Interestingly, not all sufferers of celiac disease have an autoimmune response against oats. The reason for this is not currently understood. People with celiac disease are able to eat proteins such as meats and fish, as well as fruits and vegetables. However, they must avoid many processed foods, even those that are not grain products, since many contain gluten.

The autoimmune response of celiac disease has been shown to increase the chances of getting cancers of the intestine (see Chapter 12) and lymphoma. Other problems are usually associated with malnutrition. In some cases the response to gluten will not manifest itself as intestinal problems, but as skin problems. This disease is called *dermatitis herpetiformis* and is characterized by severe itching and blisters on the buttocks, knees, and elbows.

LACTOSE INTOLERANCE

Lactose is a disaccharide sugar commonly found in milk products (see Chapter 1). It is composed of the monosaccharides glucose and galactose. The enzyme that breaks down lactose into its monosaccharide components is called *lactase.* Lactase is naturally produced by the lining of the small intestine. In the absence of lactase, the undigested lactose sugar is acted upon by bacteria in the intestinal tract, producing excessive amounts of gas (flatus). This can cause cramping, bloating, diarrhea, and abdominal pain.

Lactose intolerance is actually a natural progression of human metabolism. While some people are born with lactose intolerance, in most people the ability of the body to digest lactose naturally decreases after the age of two. The rate of decrease depends on the individual, with some people displaying symptoms of lactose intolerance early in life and others never experiencing any problems with lactose.

Lactose intolerance is easily treated by dietary means. People with lactose intolerance can either avoid milk products (the prime dietary source of lactose) entirely or use enzyme supplements to allow for lactose digestion. People who choose to avoid milk products must supplement their diet with other calcium sources, since milk products are a major source of this dietary nutrient.

DISEASES AND AILMENTS SPECIFIC TO THE LARGE INTESTINE

Appendicitis

Although the appendix is technically a part of the lymphatic system because it is comprised of lymphoid tissue, problems with the appendix are usually associated with the large or small intestine. The appendix is a small sac, located just off the cecum of the large intestine and just past the opening of the ileocecal valve. Its location makes it highly susceptible to being blocked by material entering the large intestine, especially if there is a general decrease in motility in the large intestine. Viral infections may also cause an inflammation of the appendix. This inflammation is called *appendicitis.*

The classic symptom of an inflamed appendix is pain in the right side of the abdomen. It is accented by movement, pressure, and breathing. A person may also experience nausea (see Chapter 9) and vomiting. A fever and abdominal swelling may or may not be present. It is important to note that in some cases the patient may be asymptomatic, but still have an inflamed appendix. The danger of an inflamed appendix is that is may rupture, releasing the contents of the large intestine into the body cavity and producing a severe systemic infection called *peritonitis.*

Treatment for appendicitis is usually surgery. Advances in surgical procedures have made the process, an appendectomy, a less invasive procedure. During the procedure the surgeon will make several small cuts in the abdomen and remove the inflamed appendix. Antibiotics are typically administered to reduce the chance of infection, especially if the inflammation is severe. Patients who have had an appendectomy do not need to make any modifications to their lifestyle or diet. There appear to be no long-term side effects from an appendectomy.

Constipation

Constipation is one of the intestinal ailments that is difficult to diagnose, since it represents a reduction in the number of bowel movements to the point where discomfort occurs. The average number of bowel movements varies by the individual; for some people regular bowel movements may happen daily, while others only several times per week. Some references refer to the percent of the excrement that is in the form of pellets or other hard material, yet this also varies according to the individual. A less formal but practical definition would be a reduction in the frequency or consistency of bowel movements such that it creates discomfort in an individual.

Constipation is typically associated with a slowing of movement through the large intestine. As the movement of the fecal material slows, more water is absorbed by the large intestine. Since the water in the fecal material also acts as a lubricant for movement through the large intestine, the decrease in water content further slows the movement of the stool.

There are many causes of constipation, the most common of which is a decrease in the muscle tone of the large intestine. As people age the intestinal tract naturally loses muscle tone, and a decrease in motility is therefore inevitable. However, the amount of this natural decrease can be reduced by both diet and exercise. The relationship between exercise and constipation is not well understood, but many studies have indicated that physically active people have a reduced rate of constipation. This may be indirectly linked to the fact that physically active individuals tend to have healthier diets as well.

Another common cause of constipation is due not to a problem with the gastrointestinal tract, but rather the lifestyle of the individual. People who have hectic lifestyles and repeatedly ignore the body's signals to defecate, frequently complain of constipation. Unlike the small intestine, the large intestine experiences contractions, called haustral contractions (see Chapter 3), only several times an hour. If an individual ignores these signals, then the fecal material remains in the large intestine longer, and more water is absorbed. People with hectic, high-stress lifestyles may also become dependent on laxatives to defecate, further aggravating the problem.

The most important aspect of the diet with regards to constipation is fiber. Fiber is the undigested plant material that exists in either soluble or insoluble form (see Chapter 1). As fiber moves through the gastrointestinal tract it retains water, which naturally lubricates the fecal material. In addition, it provides bulk to the food. This gives the segmentation contraction of the large intestine something to contract against, which in turn increases the motility of the large intestine. Furthermore, fiber acts as a form of resistance training for the muscles of the large intestine, meaning that is keeps them in shape and able to conduct their job. In the United States the recommended daily input of fiber is 0.88–1.06 ounces (25–30 grams) per day, but unfortunately most Americans consume around 10 grams per day.

Other factors may also play a role in constipation. Many diseases and ailments of the nervous system, including stress, influence the motility of the large intestine. Pregnancy and aging both decrease the motility of the intestinal tract, thus increasing the chance of constipation. In addition, many OTC and prescription drugs list constipation as a side-effect. People who experience constipation when beginning a drug treatment should consult with their physician before using OTC medications. The use of iron supplements, usually in women, slows the large intestine considerably and frequently causes constipation. Interestingly, the excess use of laxatives, which are medications that serve to increase motility, can cause constipation since the large intestine may become dependent on the use of medications to defecate.

Treatment for constipation depends on the severity of the problem, the length that is has been occurring, and what is known about the cause. For most people constipation is a temporary ailment that can be quickly corrected by either lifestyle or dietary changes. An awareness of the body's needs is an important first step in treating constipation. Exercise and attention to signals from the large intestine are imperative, as is a diet high in fiber with adequate water. It is important to note that most physicians and nutritionists recommend a gradual increase in daily dietary fiber intake, since sudden changes may cause bloating and gas. Some foods have natural laxative properties, including prunes, which also contain up to 6 grams of fiber per serving, and honey.

Long term, or chronic, constipation can signal a greater problem with the large intestine. Some diseases of the large intestine, such as diverticulosis or tumors, may slow movement of the fecal material. If these are suspected, your physician may use several diagnostic tools to determine the cause. One of these is a **colonoscopy**, which is sometimes called a *sigmoidoscopy* depending on the region of the large intestine being examined. During a colonoscopy, a small camera equipped with a light is inserted into the colon to look for irregularities in its structure. If a problem is encountered, small tissue samples may be removed for additional study.

A second diagnostic tool is the barium enema. In this procedure the colon is first cleaned of fecal material, then filled with a solution of barium. The barium coats the internal surface of the large intestine. This enables it to be scanned using X-rays and any irregularities in the surface structure detected. Barium enemas are relatively painless but may produce mild intestinal discomfort for several days following the procedure.

Diarrhea

The physiological opposite of constipation is diarrhea. Like constipation, it is primarily defined as a deviation in the normal bowel movements of a person over time. In diarrhea, the fecal material develops a high water content, producing more frequent bowel movements that have a high liquid content. This is usually accompanied by large amounts of intestinal gas and bloating. Diarrhea may be caused by a number of factors, including diet, infections or medications. "Traveler's diarrhea" is actually caused by an infection of the gastrointestinal tract by a virus, bacteria, or other parasite (see previous sections of this chapter). Diarrhea may also be the result of a number of other gastrointestinal ailments. All cases are of medical concern since diarrhea causes significant water loss and can lead to dehydration.

The characteristic high water content of the fecal material is due to either a decrease in absorption by the intestinal tract, specifically the large intestine, or a loss of water by the tissue into the lumen of the intestine. Acute diarrhea is usually temporary, lasting less than a few days to a week, and is usually associated with a food intolerance, food poisoning, or a bacterial or viral infection. One of the most common food intolerances is lactose intolerance (see previous section on the small intestine). Chronic diarrhea signals a greater problem with the large intestine, including inflammation (colitis), irritable bowel syndrome, celiac disease, or Crohn's disease. In addition, removal of the gall bladder may produce diarrhea as the body adjusts to its inability to store bile for fat digestion (see Chapter 11).

In most cases of acute diarrhea, the treatment for the problem involves rest and plenty of liquids. The purpose of liquids is not only to replace the lost fluids, but more importantly the lost electrolytes. Potassium is quickly lost during diarrhea, and food sources that contain potassium (such as bananas) should be added to the diet during cases of acute diarrhea. In most cases, water can be used for fluid replacement, although it does not contain sufficient electrolytes. Water or other fluids should be consumed at room temperature and on a regular basis, regardless of thirst.

Depending on the cause of the diarrhea, some medications may assist in the treatment of symptoms, although in most cases of acute diarrhea the disease must run its course. Some medications slow the movement of fecal material in the intestines by absorbing water. These commonly contain *pectins*, a soluble fiber commonly found in fruits and used in making jellies and jams, although there are other active ingredients. Another common treatment is medication containing *bismuth subsalicylate*, the active ingredient in brand name OTC medications such as Pepto-Bismol. These significantly slow the movement of material through the gastrointestinal tract.

In the case of microbial infections, some recommend the use of foods or supplements containing organisms such as *Lactobacillus acidophilus*, a bacteria naturally found in yogurt. Although their medical use is not yet approved, they appear to work at reestablishing the natural flora of bacteria in the large intestine, thus slowing the effects of pathogenic bacteria. They are not effective versus viral infections, food intolerances, or physical problems of the large intestine.

Diarrhea in any form in the elderly or small children is a potentially more dangerous disease and patients should contact their physician if symptoms persist for over 24 hours, or if accompanied by a high fever, lethargic behavior, or unresponsiveness.

Diverticular Disease

A *diverticula* is a section of the colon that has bulged outward, forming a small pouch or sac. A person who has diverticula is said to have *diverticulosis.* Waste material may accumulate in these pouches, causing an inflammation of the mucosa layer of the large intestine. Such a condition is called *diverticulitis.*

The majority of cases of diverticulosis are caused by low-fiber diets. Fiber aids in the movement of material through the lower gastrointestinal tract. Fiber also provides resistance to the muscles of the intestines, in effect allowing them to exercise and stay healthy. Without fiber, the muscalaris layers weaken over time. The FDA recommends 25 grams of fiber daily. However, most diets in western countries fall far short of that amount. Constipation, and the straining that it may cause during defecation, can also crate diverticula, especially if the colon is already susceptible due to a low-fiber diet.

Elderly people are most susceptible to diverticulosis due to the loss of muscle tone in the intestinal tract as a natural result of aging. It is estimated that the majority of people over age 60 have diverticula in the colon, although only about 25 percent of these will actually develop diverticulitis. People who have diverticulosis rarely have any symptoms, but diverticulitis is marked by pain and tenderness in the left side of the lower abdomen (although it may occur anywhere in the abdomen). Fever and nausea may be present, depending on the severity of the infection. As is the case with an appendicitis, an infection in a diverticula could spread to the surrounding tissue, causing peritonitis.

The usual treatment for diverticulitis is antibiotics and a liquid diet to give the colon a rest. Surgery is an option if the condition is severe or repeating. Dietary changes are usually strongly recommended to avoid recurrence.

Flatulence

Flatus is the medical term for intestinal gas and is a natural byproduct of our metabolism and the action of the bacteria of the intestinal tract. Gas is present along the entire length of the gastrointestinal tract. However, the intestinal gas that gives rise to flatulence, or the passing of gas through the anus, is primarily due to the action of bacteria in the colon. Excess gas in the stomach is typically expelled by burping (called **eructation**). Gas that is generated in the small intestine is typically either absorbed through the mucosal layer, or moves quickly into the large intestine.

While there are no human enzymes active in the large intestine, there is a natural flora of bacteria that reside in the colon. When undigested complex carbohydrates, such as fibers, enter into the large intestine, they are metabolized by the bacteria. Some vitamins and sugars are produced (see Chapter 3), but so is a large amount of gas. The average individual can produce between 16.5 and 15 fluid ounces (350 and 500 milliliters) of gas per day.

The production of flatus varies by individual and diet. Excessive amounts of gas in the intestinal tract can cause cramps and abdominal pain, as well as feelings of bloating in the abdomen. Excessive intestinal gas is rarely a severe medical concern unless it persists for long periods of time, in which case it may signal a more serious medical problem, including lactose intolerance. Usually the cause of large amounts of intestinal gas is dietary, and reflects a sudden change in the types of complex carbohydrates that are entering into the colon. In most cases a restoration of a normal diet will alleviate the problem.

Over-the-counter medications are available to reduce the production of intestinal gas, but no one medication can remove all gas in the gastrointestinal tract or all forms of gas. For example, OTC medications containing simethicone relieve gas buildup in the stomach, while medicines containing activated charcoal may alleviate excess gas in the large intestine. Other medications provide enzymes to help breakdown complex carbohydrates prior to their entering the large intestine. For any of these medications to be effective, the individual must first identify the food types responsible and complement the medication with dietary changes.

Hemorrhoids

Hemorrhoids are the inflammation of blood vessels in the rectum or anus. When inflamed, these blood vessels increase in size and create discomfort.

Hemorrhoids may occur internally or externally. Since they are inflamed blood vessels, often the first sign of a hemorrhoid is blood in the fecal material. External hemorrhoids may also be painful and cause an itching sensation around the anus.

There may be other causes of blood in the fecal material. However, unlike other ailments, hemorrhoids will usually clear up within a few days. A physician will normally conduct a thorough exam to rule out other diseases, such as colitis or diverticulitis.

Hemorrhoids naturally occur in people, and everyone can be expected to have a case of hemorrhoids sometimes in their lives. There are factors that increase the likelihood of developing hemorrhoids. These include persistent constipation, straining when defecating, and low fiber diets. Pregnant women frequently experience hemorrhoids due to the pressure of the fetus on the lower abdomen or the stress of childbirth.

Treatment of hemorrhoids rarely involves surgery, unless there are complications or the case is severe. Most OTC medications for hemorrhoids reduce the symptoms of the problem (such as itching), but the tissue must heal itself over a few days. Sometimes a physician may prescribe steroids to reduce the swelling of the inflamed tissue.

If hemorrhoids are linked to a dietary problem, then changes can be gradually made to decrease the chances of developing hemorrhoids in the future. Mostly this deals with restoring the fiber-water balance of the diet. The fiber content of the diet should be increased gradually (see Chapter 13) to protect against excess gas or intestinal discomfort. For people who are not able to correct fiber intake by dietary means, a physician may recommend a fiber supplement. Plenty of water (eight 8-ounce glasses of water daily) is also recommended, since this will soften the stool and prevent straining during defecation.

Hirschsprung's Disease

Hirschsprung's disease (HD) is a serious condition of the large intestine in which sections of the large intestine do not contain the nerve cells necessary for contraction. Although the disease is most often associated with the large intestine, at times the diseased sections may include portions of the ileum of the small intestine. Long-segment HD includes portions of both the small and large intestine, while short-segment HD refers to conditions involving only the large intestine. In both cases it is the result of a failure of nerve cells to develop along the length of the intestinal tract as a fetus. In other words, individuals with HD are born with the disorder.

Some cases of HD are genetic, meaning that a defective gene is inherited from the parents. HD is considered to be a complex genetic disease, meaning that a defect in any one of a number of genes may cause the disease. Researchers believe that short-segment HD and long-segment HD have different genetic causes, although both forms result in the same problems for the individual. Some forms of the disease appear to be the result of recessive mutations, while others display a **dominant** form of inheritance. However, many forms of HD are not associated with an inherited genetic disorder, and many be the result of a spontaneous mutation early in embryonic development.

As the intestines contract they naturally force material downward toward the anus. People with HD are unable to move material normally through the gastrointestinal tract. The lack of nerve cells means that the muscles of the affected section in the intestine never receive the message to contract. Thus the food moves naturally through the unaffected sections, only to stop in the affected sections. When this happens material accumulates and infections may set in. In some cases the large intestine may burst, releasing bacteria and fecal material into the abdominal cavity. This is a life-threatening situation that can cause death.

Hirschsprung's disease is normally diagnosed shortly after birth due to the inability of the newborn to have normal bowel movements. Severe constipation, sometimes alternating with diarrhea, are early symptoms. Treatment for the procedure involves surgically removing the section of the large intestine that is missing the nerve cells. This is called a *pull-through operation*, since after removing the affected section the remaining portions of the intestine are reconnected. If a significant amount of the large intestine is damaged, an ostomy may be performed. Since these procedures significantly shorten, or eliminate, the large intestine, patients must increase their consumption of water to compensate for the lack of a large intestine. If long-segment HD is diagnosed, the patient may be placed on a special diet, with supplements, to compensate for the loss of the ileum. Once the medical problems and diet have been corrected the patients can usually lead a normal life. HD is not an infection, but a developmental problem, and thus the condition can't spread to other sections of the intestine.

Irritable Bowel Syndrome

Irritable bowel syndrome (IBS) is an ailment that is best characterized as a chronic irregularity in bowel movements. These irregularities may involve constipation, diarrhea, or a constant fluctuation between the two. The symptoms of IBS vary by the individual. Unlike many of the ailments of the large intestine mentioned in this chapter, IBS is more of a nuisance than a major medical problem.

It is unclear as to what actually causes IBS. What is known is that IBS is a separate disease from all of the other ailments of the large intestine, including colitis, Crohn's disease, and diverticulitis. In fact, it is usually only after a physician has concluded that there are no physical problems with the large intestine that IBS is diagnosed. Currently, there is no cure for IBS, although the symptoms may be treated to make the patient more comfortable.

While the cause of IBS is also unknown, two factors are known to influence the severity of the symptoms. The first is stress. Stress is a physiological response by the body to an outside stimulus, and is well known to create problems with both diarrhea and constipation. In people suffering from IBS, the effects of stress are more pronounced, resulting in more severe symptoms or rapid fluctuations between constipation and diarrhea. The second factor is diet. People with IBS respond to the stimulants in foods, such as caffeine, in different ways. In all cases, stimulants cause the muscular lining of the large intestine to go into spasms, or uncontrolled contractions. This may lead to diarrhea or constipation, depending on the individual. Fats and oils are also commonly a factor, since the presence of these in the gastrointestinal tract increases the rate of contraction of the muscles in the large intestine.

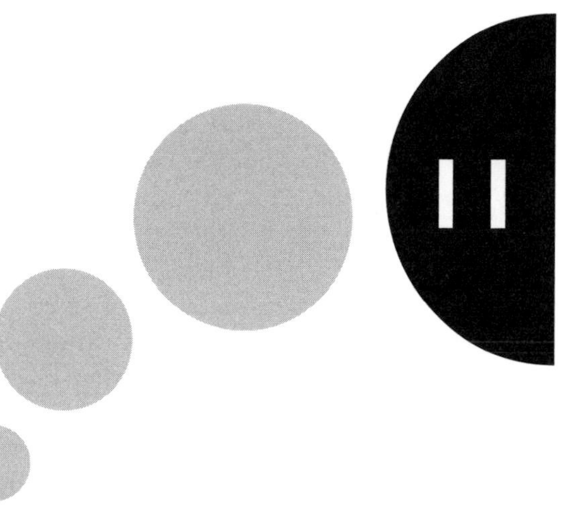

Diseases and Ailments of the Accessory Glands

While the accessory glands do not directly come in contact with the food, the secretions of these glands are crucial (with the possible exception of the gall bladder) to the efficient operation of the digestive system. In addition, the liver and pancreas are multi-function organs that interact with a number of systems in the human body. For example, the pancreas is both a digestive organ and an endocrine organ, while the liver also interacts with the circulatory system. Since the nutrients provided by the digestive tract are necessary for the efficient operation of all cells in the body, problems with the accessory glands can have a pronounced effect on human health and physiology.

This chapter examines some of the more common ailments and diseases of the accessory glands and focuses on those that are related to the health of the digestive system. Cancer, a disease that affects all of the organs of the human body, is covered in detail in Chapter 12.

SALIVARY GLANDS

The primary purpose of the salivary glands is to moisten the oral cavity and esophagus for the efficient transfer of food to the stomach. Thus it should not be of any real surprise that problems with the salivary glands tend to result in symptoms of dry mouth. There are several different ailments that may be associated with a dry mouth, the two most common are *xerostomia* and *Sjögren's syndrome.*

Xerostomia is the general medical term for dry mouth. It is not itself a disease, but usually a symptom of another medical problem. Simply put, xerostomia is the result of a reduction in the saliva output by the salivary glands. This may simply be a result of the thirst response by the body or a sign of dehydration. If this is the case then this temporary condition can be resolved by simply rehydrating the body by the intake of water. Xerostomia is also common with certain medications and is usually indicated as a side-effect with the literature provided with the drug. Anti-depressants, medications to treat **hypertension**, and **diuretics** commonly cause xerostomia, although the list is much more extensive than this. Since each person possesses unique variations in their biochemistry and metabolism, patients who experience xerostomia when beginning a new medication regime should consult with their physicians.

Xerostomia may also be caused by stress or injury. Since the production of saliva is a result of a reflex response between the sensory inputs of sight and smell and the medulla of the brain stem, any injury which interferes with the relaying of signals between the brain and salivary glands may drastically reduce saliva production. Patients who suffer from **strokes** or nerve damage in the neck area may show symptoms of xerostomia. In stressful situations, the sympathetic nervous system dominates and the resulting fight-or-flight response, a natural action of the nervous system, decreases saliva production. In fact, a dry mouth is a common sign of stress and anxiety.

Scientists have recently determined that aging does not directly affect the ability of the salivary glands to produce saliva. Instead, a natural decrease in the ability of the person to detect sensory input, combined with medication, inhibits the saliva reflex and decreases saliva production.

Xerostomia is not simply an inconvenience for the person who experiences it. The production of saliva is an important mechanism for protecting the oral cavity from bacteria and injury. Without saliva, the chemistry of the oral cavity is frequently altered, allowing more rapid bacterial growth. In addition, the interior epithelial linings of the mouth may crack without saliva, producing painful cuts and increasing the chance of infection.

Treatment for xerostomia varies with the cause. Adequate hydration is crucial not only to provide enough water for saliva production, but to continuously moisten the oral cavity. Patients with xerostomia should limit the use of any diuretics, including caffeine. Most physicians recommend that the patient use sugarless hard candies or gum to increase sensory stimulation in the oral cavity. In some cases, artificial salivas may be prescribed.

A dry mouth may also be an indication of a disease called Sjögren's syndrome. The disease is named after the Swedish ophthalmologist Henrick Sjögren, who first described the condition in 1933. Sjögren's syndrome has a wide variety of symptoms, including dry mouth, dry eyes, fatigue, and joint pain. Sjögren's syndrome is an autoimmune disease in which the immune system of the body mistakenly targets the moisture producing glands, including the salivary glands. This type of condition is called *primary Sjögren's*. Women are nine times more likely to get this disease than men and the disease is rarely seen in people under forty years of age. While the symptoms of this disease can be treated, the concern is that the disease may also affect other glands of the body, such as the liver, skin, reproductive glands, and nerve cells. Sjögren's syndrome has been closely linked to liver problems such as primary biliary cirrhosis, cirrhosis, and hepatitis (see following section on the liver).

The symptoms of Sjögren's syndrome may also be the result of a connective tissue disease, such as rheumatoid arthritis (see Muscular System volume of this series). This is called *secondary Sjögren's syndrome*. What is interesting is that the primary Sjögren's rarely results in secondary Sjögren's, but secondary Sjögren's frequently causes primary Sjögren's. The cause of Sjögren's is not known, but it believed to begin after an unidentified viral infection. In a small percent of the cases secondary Sjögren's can lead to a form of cancer called **lymphoma** (see Lymphatic System volume of this series). Little is known about this disease, and research is currently underway to determine the various causes.

Cystic Fibrosis and Salivary Function

Another disease that may influence the operation of the salivary glands is cystic fibrosis (CF). Cystic fibrosis is a disease of epithelial cells. It is believed that a defect in the transport of salt ions, specifically sodium and chloride, across the cell membranes of CF cells is to blame. If the cells are excretory cells, such as the salivary glands and pancreas, then the fluid concentration of the secretions is typically lower, creating a much more concentrated secretion. Some researchers believe that the faulty ion transport is at the organelle level within the cell, and not only in the cell's outer membranes.

In the case of the salivary glands, the decrease in water content results in thicker mucous secretions by the glands. This increases the chance that the ducts connecting the glands to the oral cavity may become blocked. Since the mouth is home to a multitude of bacteria, a blocked duct represents an ideal location for a bacterial infection to occur.

While CF is primarily a disease of the respiratory system (see Respiratory System volume of this series), scientists are studying the fluid-regulating pathways of salivary glands to better understand how CF may interfere with fluid regulation in other epithelial cells. Their research has primarily focused on the signaling mechanisms by which these cells are activated. With a greater understanding of this process, it may be possible in the future to artificially stimulate epithelial cells in CF patients to secrete the proper water content.

Cystic fibrosis is a **recessive** genetic disorder, meaning that an individual can carry the trait and yet show no outward signs of the disease. Thus, both parents must contribute a defective copy of the gene to their offspring in order for the child to have CF. If both parents carry the defective trait, there is a 25 percent chance that their offspring will have CF. Over 30,000 children are currently diagnosed with CF, and even with treatment few will live past age 30. There is no known cure, although modern research is investigating using **gene therapy** to correct the defective gene, or introduce a good copy, in an infected person.

Mumps

Mumps is a common childhood disease that is caused by a virus called a *paramyxovirus*. This virus infects a number of tissues in the body, but is most often associated with the parotid salivary glands. Since the parotids are located just below the ears, an infection of these glands results in a difficulty chewing and a swelling just under the ears. In 25 percent of all cases, no symptoms may be present. There are cases where the other salivary glands may be infected. In small children, the disease is usually accompanied by a fever. In males, there may also be soreness and swelling of the testicles, called *orchitis*. In some cases this may result in decreased sperm count or infertility. In rare cases, the disease may cause **encephalitis** or **meningitis**.

Since mumps is a disease of the salivary glands, it is easily transmitted by coming in contact with the saliva of a infected person. Sneezing and coughing each vaporize saliva, thus increasing the chances that the virus will be spread. There is no treatment for mumps other than rest. Since this is a viral disease, antibiotics are ineffective. The disease usually runs its course within fourteen days.

A vaccination against mumps, called the MMR vaccine (for measles, mumps, and rubella) is commonly first given to children between 12 and 15 months of age, with a second dose between the ages of 3 and 6 years old. Recently there has been some controversy regarding this vaccine as some claim that the MMR vaccine may be associated with an increased rate of **autism**. However, extensive analysis by the Centers for Disease Control and other government organizations have failed to indicate any link between the two. The chances of permanent harm to an individual as a result of mumps, measles, or rubella far exceeds the undocumented risk of autism. Since the vaccine was first introduced in 1967, the rate of mumps has decreased in the United States from 200,000 cases per year to less than 1,000.

A common misconception regarding mumps is the notion that people can get the disease more than once. Since mumps may not produce symptoms in both parotid glands, it is possible for a sufferer to believe that he or she only got mumps on one side, and thus is still susceptible to get it in the other side. However, since mumps is the result of a viral infection, the immune system generates **antibodies** against the virus, which effectively prohibits a second infection. People who claim to have had mumps twice most likely experienced a separate infection of the parotid glands by an agent other than mumps.

Waterbrash

Either just prior to, or during regurgitation, the salivary glands may participate in a reflex reaction called *waterbrash*. In response to gastric juice in the esophagus, the salivary glands may increase secretion. These secretions tend to be clear and have a high salt concentration. Waterbrash is not the same thing as gastroesophageal reflux disease (GERD; see Chapter 8), although its symptoms may seem similar to the patient and the two may occur simultaneously. Waterbrash is sometimes also called *hypersalivation*. Persistent occurrences of waterbrash does not produce any physiological problems for the individual, although it may be a symptom of other medical problems, and the condition should be discussed with a physician.

GALL BLADDER

Gallstones

The purpose of the gall bladder is to store bile, a compound produced by the liver (see Chapter 4) that is utilized in the small intestine for fat digestion (see Chapter 3). Bile is comprised of bile salts, water, cholesterol, and bilirubin. The liver is connected to the small intestine by the hepatic duct. Branching off from the hepatic duct is the cystic duct, which leads to the gall bladder. Until bile is needed by the small intestine, it backs up and is stored within the gall bladder (see Chapter 4). The movement of the bile within these ducts requires that the correct proportion of cholesterol, bile salts, bilirubin, and water be present. If there is an imbalance in these proportions, then the bile may harden into small masses called *gallstones*.

There are two general types of gallstones. *Pigment stones*, the rarer of the two, are produced by excess bilirubin. These stones are dark in color and are a sign of a potential problem with the liver or circulatory system. The more common gallstone is the *cholesterol stone*, which are a greenish-yellow in color. These are associated with an excess of cholesterol in the bile, or too little water. Gallstones may develop in a variety of sizes and shapes. Although these types of gallstones have different origins, their effects on the body are the same.

The primary medical problem associated with gallstones occurs when they block one of the ducts between the liver, gall bladder, or pancreas and the small intestine. Depending on the location of the blockage (see Figure

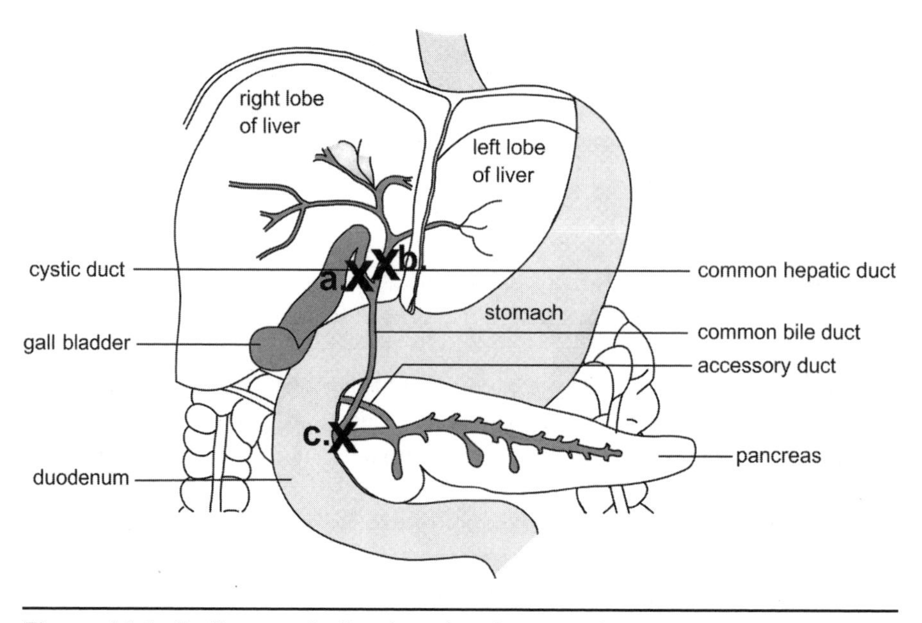

Figure 11.1. A diagram indicating the three major locations where gallstones may occur.
The symptoms of a gallstone are dependent on its location.

11.1), the secretions of the accessory organ are unable to reach the small intestine and back up within the organ. Since the pancreas shares a segment of the bile duct, a blockage at this point may cause digestive enzymes to accumulate in the pancreas, causing a disease called *pancreatitis* (see later section). If the blockage occurs in the cystic duct, the tissues of the gall bladder may become inflamed or infected. A blockage in the hepatic duct can cause an inflammation of the liver.

Anyone can develop gallstones, but it is more prevalent in obese people or those suffering from liver problems. Many medications that lower blood cholesterol increase cholesterol secretion by the liver, thus increasing the amount of cholesterol in the gall bladder and upsetting the chemical balance of the bile. The use of other medications, including birth control pills (which are primarily the hormone estrogen) or hormone replacement therapy, may also increase the chances of developing gallstones. In general, women are more susceptible then men to developing gallstones.

Many individuals who develop gallstones have no outward symptoms. This is especially true if the stones are small and do not block any major ducts. Other symptoms include abdominal pain, especially in the upper right areas of the abdomen (the location of the gall bladder) or frequent problems with indigestion or nausea. If the gallstone completely blocks the secretions of the liver, a person may develop *jaundice*. Jaundice is characterized by a yellowing of the skin and is a severe medical condition that requires immediate treatment.

Treatment for gallstones depends on the individual, and the location and size of the gallstone. Surgery to remove the entire gall bladder, called a *cholecystectomy* (see next section) or to remove the gallstone may be performed. Although the gall bladder is not necessary for human health, increasingly physicians are focusing on non-surgical options to remove gallstones. This may include the use of medications to dissolve the gallstone in place. These medications may be taken orally or injected directly into the gall bladder. In some cases, sound waves may be used to break the gallstone into smaller pieces so that it can pass harmlessly into the small intestine and be secreted. This procedure is called *extracorporeal shockwave lithotropsy*, but its success depends on the composition and size of the original gallstone.

Cholecystectomy

A cholecystectomy is a surgical procedure to remove an infected gall bladder. Unlike most of the organs of the digestive tract, a gall bladder is not necessary for digestion, and thus individuals without gall bladders can live completely normal lives. The gall bladder may be removed for a variety of reasons, the most common of which being that is has become diseased or inflamed by gallstones.

The most common surgical procedure to remove a gall bladder is called a *laparoscopic cholecystectomy*. In this procedure a small incision is made in the abdomen and miniature instruments inserted into the body cavity. Along with the instruments, a small camera is inserted so that the surgeon can visually monitor the procedure on a video screen. The operation involves the removal of the gall bladder and cystic ducts, and the sealing of the hepatic bile duct. Following the procedure, bile proceeds directly to the small intestine without being diverted into the cystic duct and gallbladder. A large incision is sometimes made in the abdomen, but the more common procedure uses less invasive laparoscopic technology.

In general, individuals who have had their gall bladder removed do not need to restrict fat intake or make other changes to their diets. They simply lack a bile storage location. This means that bile is being continuously supplied to the small intestine. This may cause diarrhea in some individuals, but most people do not experience any negative side effects.

LIVER

The liver plays a central role in nutrient metabolism and works closely with a number of other body systems, including the circulatory, lymphatic and endocrine systems. There are many diseases that may affect the operation of the liver; this chapter will examine those that are primarily associated with diet or digestion.

Cirrhosis

Cirrhosis of the liver is sometimes also called hardening of the liver, and is a disease generally associated with the abuse of alcohol. Alcohol is readily absorbed through the mucosal layer of the gastrointestinal tract, and quickly makes its way via the portal circulatory system (see Chapter 4) to the liver. The hepatocytes in the liver are responsible not only for the formation of bile and the sequestering of some nutrients, such as iron, but also the detoxification of harmful substances in the food. When alcohol is present, the hepatocytes cease nutrient processing and focus on detoxification of the alcohol. As a result, fatty acids accumulate in the liver tissues, causing a condition known as *fatty liver*. The fat in the liver disrupts all other functions, effectively rendering the liver tissue unable to conduct its normal tasks. If the liver is not able to resume its normal function due to the continuous abuse of alcohol, then the fatty deposits are replaced by hardened scar tissue. The damage at this point is usually irreversible, although the progression of the disease may be halted.

Other factors may contribute to the development of cirrhosis besides alcohol use. Hepatitis, a disease caused by a viral infection (see later section), may cause cirrhosis, as can other systemic infections or even an autoimmune response against the cells of the liver. Due to the redundant structure and regenerative nature of the liver (see Chapter 4), the symptoms of cirrhosis may take years to present themselves. Early symptoms may include nausea, fatigue, weight loss, and loss of appetite. As the disease progresses, other symptoms may develop, depending on what functions of the liver are influenced by the replacement of healthy tissue with scar tissue. For example, if the scar tissue limits the movement of bile, gallstones may develop in the liver. An interference with the processing of bilirubin may result in jaundice. Since the liver functions as the prime location of detoxification in the body, any loss of function may result in increased sensitivity to medications or an accumulation of metabolic toxins in the blood. In the later case, other organ systems may also be affected. Cirrhosis can be fatal; the disease is responsible for over 25,000 deaths annually.

Hemochromatosis

Hemochromatosis is also sometimes called *iron overload*. In this condition, too much iron is absorbed from the small intestine. While iron is frequently viewed as a beneficial nutrient, and most problems are associated with deficiencies, an excess of iron is toxic to the tissues of the liver. Iron-rich tissues also are susceptible to bacterial infections, thus enhancing the risk of excessive iron absorption.

Several different factors may cause hemochromatosis. In a large number of cases, the condition is hereditary. Excessive use of alcohol also increases the risk of hemochromatosis, because alcohol damages the body's ability to regulate iron absorption across the microvilli of the small intestine. Use of high-iron supplements and blood transfusions also increase the risk of hemochromatosis.

Signs of hemochromatosis parallel those of anemia. Frequently the individual feels very lethargic and may be very apathetic. It is often difficult to distinguish between excessive and deficient iron states. Treatment involves a strict regulation of the diet, including a reduction in iron-enriched or iron-fortified foods.

Hepatitis

Hepatitis is an infection of the liver caused by a family of viruses. It is one of the most common liver ailments in the United States, with some estimates by the Centers for Disease Control (CDC) suggesting that more than 5 million American may be infected. Currently there are six known viruses that cause hepatitis; these are listed in Table 11.1 along with a summary of their major forms of transmission.

There are two major forms of hepatitis infections: acute and chronic. Acute infections last a short period of time, followed by either a remission or recovery by the patient. Of the forms of hepatitis listed in Table 11.1, hepatitis A is an example of an acute form. In a chronic condition, the infection is ongoing and may result in long-term liver damage, especially in those individuals with hepatitis B or C. This may include cirrhosis of the liver or liver cancer (see Chapter 12). In some cases of chronic hepatitis, the individual may be asymptomatic but still carry the virus and transmit it to others. This is especially true of hepatitis C, but also occurs in some cases of hepatitis B.

Symptoms of a hepatitis infection include fatigue, a yellowing of the eyes, dark urine, tenderness of the abdominal area, and weight loss. The darkening of the urine, which is sometimes accompanied by a lightening of the stool, and yellowness of the eyes indicates the interference of normal liver function by the virus. While vaccinations may protect against certain forms of the virus, there are no remedies for a hepatitis infection, although some antiviral medications under development may show promise.

Protection against a hepatitis infection focuses on awareness. Individuals traveling to countries with questionable sanitation procedures are advised to get immunized against hepatitis A. The U.S. State Department maintains a current list of those countries for which immunizations are recommended. While in these countries, individuals are advised to avoid drinking unpurified water or eating shellfish. Other forms of hepatitis are transmitted sexually or through the shared use of intravenous needles for drugs. It is not

TABLE 11.1. Forms of Hepatitis

Form	Transmitted Via*	Vaccination Available	Long-term Liver Damage	Notes
Hepatitis A	• Fecal contaminated water supplies • Sexual contact • Consumption of shellfish from fecal contaminated water	Yes	No	
Hepatitis B	• Sexual contact • Contact with body fluids of infected person • Intravenous drug use	Yes	Yes	
Hepatitis C	• Blood • Intravenous drug use	No	Yes	• May cause cirrhosis or liver cancer
Hepatitis D	• Sexual contact • Contact with body fluids of infected person • Intravenous drug use	No	Yes	• Must be present with hepatitis B • More rapid deterioration of liver
Hepatitis E	• Contaminated food or water supplies	No	No	
Hepatitis G	• Blood and blood transfusions	No	Unknown	• Discovered in 1995

**Indicates the most common forms of transmission*

uncommon for some individuals to have multiple forms of hepatitis, which in many cases greatly accelerates progression of liver damage.

PANCREAS

Pancreatitis

Pancreatitis is defined as an inflammation of the tissues of the pancreas. Since the pancreas is both an endocrine and exocrine organ, any damage to its tissues may present a severe problem for the individual. Close to 3,000 people die annually from pancreatitis, and close to one million cases are diagnosed each year.

An inflammation of the pancreatic tissues may be the result of a number of factors, the most common being the abuse of alcoholic beverages and blockage of the pancreatic ducts due to gallstones. Some cases have been reported due to traumas to the abdomen or pancreatic tumors.

The symptoms of pancreatitis include abdominal pain or a tender and swollen abdomen. Some patients may experience nausea, vomiting, and fever as well. The disease has two distinct forms. The acute form is temporary and usually lasts a few days before resolving itself. A more significant medical problem is chronic pancreatitis. In this form of the disease, the digestive enzymes secreted by the pancreas, which are normally inactive (see Chapter 4), begin to slowly digest the tissues of the pancreas, causing extensive tissue damage and inflammation.

Acute pancreatitis is usually diagnosed using a blood test that looks for abnormal levels of pancreatic enzymes. The causes of acute pancreatitis are not always known, but the use of alcohol can cause the disease, as can traumas to the pancreas. In some cases, cysts, or fluid-filled tissues, interfere with the normal operation of the pancreas. Gallstones may inhibit the excretion of pancreatic enzymes, causing pancreatitis until the gallstones are removed. In general, treatment for acute pancreatitis involves allowing the tissues of the pancreas to rest, with frequent monitoring to ensure that the condition is not spreading. Patients may be given intravenous fluids and placed on a restricted diet until the tissue heals.

Chronic pancreatitis represents a much more severe medical threat, one that may require years to manifest itself. Chronic pancreatitis is almost always associated with alcoholism, although there are hereditary forms of the disease that affect some families. Symptoms of chronic pancreatitis mirror those of the acute form, although they are recurring and frequently more severe. In some cases, the patient eventually does not experience pain as the pancreatic tissue becomes so damaged it stops enzyme production. In many cases of chronic pancreatitis, the individual experiences weight loss while maintaining an adequate diet. This is due to the lack of pancreatic digestive enzymes in the small intestine and the resulting failure of the body to process nutrients. In all cases of chronic pancreatitis, the tissues of the pancreas may be permanently destroyed and replaced by scar tissue.

Treatment for chronic pancreatitis involves not only allowing the pancreatic tissues time to rest, but also addressing the cause of the disease. If the root is alcohol abuse, the individual must abstain from alcohol use, which may be difficult for individuals suffering from alcoholism. In many cases, a physician may prescribe a hospital stay with nutrients and fluids delivered intravenously. Antibiotics may be prescribed to inhibit secondary infections. Digestive enzymes may also be supplied to ensure that the body is properly metabolizing nutrients for healing to occur. Chronic pancreatitis can sometimes lead to the permanent destruction of the endocrine regions of the pancreas, resulting in **diabetes**.

Zollinger-Ellison Syndrome

Zollinger-Ellison syndrome is a relatively rare disease of the pancreas that usually manifests itself as problems with the duodenum or stomach. In Zollinger-Ellison syndrome, pancreatic tumors induce the pancreas to secrete gastrin, a hormone that causes the stomach to increase gastric juice production. The result is hyperacidity of the stomach, which usually results in gastritis, gastric ulcers, or duodenal ulcers. Treatment for Zollinger-Ellison syndrome may involve surgical removal of the tumors, but often simply involves using medications to treat the increased acid secretions of the stomach.

Cancer of the Digestive System

Cancer is characterized as the abnormal, uncontrolled growth of cells. It represents a change in the genetic programming of the cell so that the cell now disregards external and internal regulatory signals. The change in the genetic programming may the result of a *mutagen*, a chemical that causes a change in the DNA, or the activity of a virus or a **transposon**. If the mutagen causes cancer in the cell, it is called a *carcinogen*. Both natural and manmade compounds may be carcinogens, as are many forms of radiation.

Cancers are named for where they originate in the body. This site also denotes the *primary tumor*. For example, pancreatic cancer originates from a tumor of the pancreas. In some cases, cancer cells may migrate from the primary tumor to another organ of the body. This process is called **metastasis** and causes the formation of secondary tumors. It is important to note that the secondary tumor contains cell types from the original tumor. If the primary tumor can't be identified, the cancer is called a *carcinoma of unknown primary*, or CUP.

As cells grow they form tumors, which are a mass of cells that usually originate from the original cell. Not all tumors are cancers. A *benign tumor* does not invade the surrounding tissue and does not undergo metastatic activity. Tumors that actively invade tissues and spread around the body are called *malignant tumors*. Benign tumors may become malignant over time.

The cellular basis of cancer is complex, and researchers are just beginning to decipher the various mechanisms by which a cell becomes a cancer cell. Basically, all cells undergo a regular cell cycle of division. This cell cycle involves a replication of the genetic material and a series of checkpoints that must be passed before the cell can undergo division. The checkpoints examine the integrity of the DNA, environmental factors, cell density, and a variety of other factors. If a mutation in the genetic material enables the cell to proceed past a checkpoint that normally would be closed, the cell may begin unregulated cell division. Researchers have identified an array of genes and biochemical mechanisms that may enable a cell to ignore the checkpoints. Since cancer represents one of the leading causes of death, cancer research represents a major area of activity, with scores of geneticists, chemists, and biochemists working on the problem around the world.

The material provided in this chapter is designed to provide a general overview of cancer types associated with the organs of the digestive system (see Table 12.1). The field of cancer research is rapidly evolving and represents one of the main focuses of medical research. New advances are an-

TABLE 12.1. Summary of Digestive Cancer Types

Location of Cancer	Primary Cause	Symptoms*	Five-Year Survivor Rate**
Oral	Tobacco use; alcohol use	Lumps or sores in the oral cavity that do not heal	90%
Esophageal	Tobacco use; alcohol use	Difficulty swallowing	50%
Stomach	Unknown; may be related to food additives or *H. pylori*	Abdominal pain, vomiting, and unexplained loss of weight	45%
Colorectal	Environmental contaminants coupled to a high-fat, low-fiber diet	Persistent diarrhea or constipation, abdominal pain, blood in the feces	90%
Liver	Hepatitis, cirrhosis, industrial chemicals	Abdominal pain, weight loss	<30%
Pancreatic	Unknown, possibly linked to pancreatitis	Digestive problems, weight loss, pancreatitis	15%
Gall bladder	Unknown, possibly linked to gallstones	Jaundice, abdominal pain and swelling	80%

** A few primary symptoms; symptoms may vary depending on location and individual*
*** Average survivor rate for a stage I cancer.*

nounced almost daily, and scientists are rapidly closing in on the causes of most cancers. A list of cancer information sites is provided in the Organizations and Web Sites chapter at the end of this book. The resources listed there contain information on recent advances and treatments, as well as contact information on cancer specialists and support groups for individuals with cancer and their families.

STAGING OF CANCER

Medical professionals have developed a system by which they classify tumors. This process is called *staging*. There are several forms of staging, but the most common is called the *TNM system* developed by the American Joint Commission on Cancer. The TNM system identifies three aspects of a tumor:

- T represents the size of the tumor and if it has become invasive of surrounding tissues.
- N represents the extent to which a tumor has begun to invade the surrounding lymph nodes.
- M represents where the tumor has begun to metastasize and spread to other organs of the body, forming secondary tumors.

Each of these letters is assigned a number between 0 and 5. The number indicates the degree of each category, with 0 being the lowest. Thus a T1 N0 M0 tumor would be much less threatening than a T3 N2 M2 tumor. The goal of physicians is to diagnose cancers at the earliest possible stage, since this then gives them the greatest number of options for treatment, and the greatest chance of success.

Each form of cancer has minor variations in the staging of the disease to reflect the organ or tissue in which it originated. The information provided in this chapter is designed to provide a general reference. Cancer patients should feel free to request additional information from their physician.

CANCERS OF THE DIGESTIVE TRACT

Oral Cancer

Oral cancer is a general term that is assigned to any cancer of the mouth, pharynx, or salivary glands. In almost 90 percent of the cases of oral cancer, the individual has a history of tobacco use. This includes the use of chewing tobacco, cigarettes, and pipes. The use of alcoholic beverages increases the risk of developing cancer of the oral cavity, especially if combined with tobacco use. Because cancer is associated with abnormal cell growth, and the oral cavity contains a wide variety of different cell types, there may be other environmental influences that can cause oral cancer in non-smokers or those who do not use alcohol. However, cancer of the oral cavity is rare in these individuals.

Oral cancers almost always originate in the linings of the oral cavities. These are commonly called *squamous cell carcinomas*. Occasionally the cancer may originate in the salivary glands, although cancer of the salivary gland is not usually linked to alcohol or tobacco use, but rather exposure to radiation as a child.

The symptoms of oral cancer include lumps in the cheek or sores in the mouth that do not heal after a few weeks. A patchy appearance on the gums, tongue, or internal linings of the cheek may also indicate a problem. These patches may be either red or white. Other symptoms include difficulty in swallowing or in the movement of the jaw. Salivary gland cancer includes swollen glands and sometimes pain in the facial nerves. These symptoms may be caused by a number of other factors, but if they exist they should be brought to the attention of a physician. Oral cancer may quickly spread to other tissues of the body, especially the lymphatic system, if not detected early. The outcome is usually more favorable through early detection. The five-year survival rate for a stage I cancer is over 90 percent, depending on the location of the cancer.

Esophageal Cancer

Like oral cancer, esophageal cancer is usually the result of tobacco use or abuse of alcoholic beverages. Esophageal cancer is otherwise rare in the United States. Some studies have linked diets high in nitrates and nitrites as a risk factor for the development of esophageal cancer. In the body these chemicals are metabolized into a chemical called a *nitrosamine*, which has been identified as a carcinogen. As was the case with oral cancer, the majority of esophageal cancers form in the linings of the esophagus, indicating the environmental influence on this form of cancer.

Symptoms of esophageal cancer include pain or difficulty swallowing, a persistent cough usually accompanied by blood, and hoarseness or difficulty in talking. There are other diseases that may cause these symptoms. However, the presence of any of these symptoms should be brought to the attention of your physician. The primary problem of esophageal cancer is that by the time the symptoms manifest themselves, the cancer has usually progressed past stage I, when treatment is most effective. Unfortunately, survival rates for esophageal cancer are relatively low. The five-year survivor rate for a stage I esophageal cancer is only 50 percent, with rates decreasing as the cancer progresses.

Stomach Cancer

Stomach cancer is another of the rare forms of cancer that appears to be decreasing in occurrence in the United States. The exact causes of stomach cancer (sometimes called *gastric cancer*) are not clearly understood. Some studies have linked food additives, mostly nitrates and nitrites, to an increased risk, while other studies have implicated the stomach parasite *Heliobacter pylori* (see Chapter 10) as increasing the risk of stomach cancer. If the latter is the case, then current success in identifying and treating *H. pylori* infections should further reduce the rate of stomach cancer in the future.

Stomach cancer usually forms in the mucosa layer of the stomach, and usually in the lower regions of the stomach near the pyloric sphincter. The primary problem with stomach cancer is that it is usually not diagnosed in the first stages. This is because the symptoms of stomach cancer closely mimic those of other stomach disorders, most notably peptic ulcers. Stomach cancer produces pain in the abdomen, but this pain is usually not alleviated by eating, as is the case with an ulcer. Vomiting, blood in the feces, unexplained weight loss, or fluid accumulation in the abdominal cavity are all signs of stomach cancer. If diagnosed early, the five-year survival rate for a stage I cancer of the stomach is around 45 percent.

Colorectal Cancer

The most common location of cancer in the intestines is in the colon. However, there are rare forms of cancer that may affect the small intestine as well. Colorectal cancer is one of the more common human cancers and represents one in which the occurrence in the population is increasing over time. The colon primarily handles undigested and unabsorbed material from the small intestine. In this capacity, the cells of the colon are frequently in contact with environmental contaminants, such as pesticides and heavy metals that are found in our food supply. In addition, undigested animal fat often makes its way into the large intestine, where it may be broken down into carcinogenic compounds by intestinal bacteria. Some studies have suggested that bile, a natural product of the liver, may also be carcinogenic when metabolized by intestinal bacteria. As a result of the exposure to these compounds, the lining of the colon forms abnormal growths of tissue, called *polyps*. These polyps are believed to be the precursor of colorectal cancer, although not in all cases will polyps become malignant.

The symptoms of colorectal cancer include blood in the fecal material. Because the tumors may block a portion of the colon or rectum, there may be a feeling that the bowel has not completely emptied following defecation. The stool may also be narrower than normal, or the individual may have to strain to complete a bowel movement. Abdominal pain may be present, as may be a bloated feeling in the abdominal cavity. The use of annual exams for individuals over age 40 has increased the chances that colorectal cancer is diagnosed early. The survival rate for a stage I colorectal cancer is 90 percent, although this percentage is increasing as individuals undergo regular screening and the public is better informed.

Liver Cancer

Cancer of the liver is a very rare but very serious condition due to the importance of the liver to both the digestive and circulatory systems. Liver cancer is rarely a primary cancer, but usually represents the metastatic activity of a cancer somewhere else in the body. Most cases of primary liver cancer are the result of viral infections, namely hepatitis B and C (see Chapter 11). Cirrhosis of the liver, usually caused by alcohol abuse, has been implicated in the development of liver cancer, as has exposure to some industrial chemicals and environmental contaminants. A fungal toxin called *aflatoxin* has also been shown to be a carcinogen of the liver.

Symptoms of liver cancer include a change in appetite, pain in the abdominal area (usually on the right side of the abdominal cavity), and a lump below the ribs (also usually on the right side). Jaundice and unexplained weight loss may also occur, and there may be an accumulation of fluids in the abdominal cavity. If the cancer is secondary, these symptoms may or may not be accompanied by the symptoms of the primary cancer. Surgery is often the option, since the liver has the ability to regenerate over 80 percent of its mass. Liver tumors that are eligible for surgery are called *resectable*. The five-year survival rate for resectable tumors is around 30 percent, but falls to less than 1 percent if there are no surgical options.

Pancreatic Cancer

The pancreas is both an endocrine and an exocrine gland, and thus ailments of this organ are covered in more detail in the Endocrine System volume of this series. However, because the pancreas plays such a crucial role in the operation of the large intestine, pancreatic cancer will be presented in this chapter.

Pancreatic cancer is ranked as the second most common type of gastrointestinal cancer, following colorectal, and the ninth most frequent form of cancer in humans. The cause of pancreatic cancer is not completely understood, but those with a history of pancreatitis (see Chapter 11) have a higher rate of pancreatic cancer. There may also be a genetic connection, because pancreatic cancer displays inheritance patterns suggesting an increase in risk factors in certain populations. Industrial chemicals and environmental contaminants may also increase the risk of developing pancreatic cancer.

The symptoms of pancreatic cancer are dependent on which area of the pancreas is being affected by the tumor. If the tumor is in the exocrine areas of the pancreas, which supply the digestive tract with enzymes, then there is usually a loss of appetite and unexplained loss of weight. The person may experience difficulty digesting certain foods, usually those with a high fat content. They may also experience pancreatitis, a painful inflammation of the pancreas. Abdominal pain is common, as is the development of diabetes. There is often a change in the coloration of the body's wastes, with the urine becoming very dark and the stool very light. The five-year survival rate for a stage I pancreatic cancer is only 15 percent.

Gall Bladder Cancer

Cancer of the gall bladder and bile ducts represents another rare form of cancer, one that is usually not diagnosed until other problems develop with the gall bladder (see Chapter 11). The function of the gall bladder is to store bile from the liver. In some people, this bile hardens to form gallstones. The presence of gallstones may irritate the internal lining of the gall bladder, making it more prone to cancer development. The presence of gallstones does not indicate that cancer is inevitable, but the majority of people with cancer of the gall bladder have a history of gallstones.

The symptoms of gall bladder cancer include jaundice, nausea, and abdominal pain or swelling. This may be accompanied by an unexplained loss of weight and a lump in the abdomen. Dark colored urine may also be present, because bile is responsible for the coloration of the urine. Survival of gall bladder cancer depends on whether or not the tumor can be removed. If it can, usually by removing the entire gall bladder (see Chapter 11), then the survival rate is in excess of 80 percent. If the cancer has spread into the liver, which is a definite possibility due to the close proximity of the gall bladder to the liver, then the five-year survival rate is lowered to less than 5 percent.

CANCER TREATMENTS

The treatment options for a specific cancer are dependent on a variety of factors:

- Health of the individual
- Stage of the cancer
- Aggressive nature of the cancer
- Location of the cancer
- Availability of treatment

- Side effects of the treatment
- Cost of the treatment
- Individual preferences

Once the condition of the patient and the extent and location of the cancer have been determined, a treatment regime will be developed by a physician. Physicians that specialize in the diagnosis and treatment of cancer are called *oncologists*. This section discusses the more common cancer treatments, but new innovative methods of treating cancer are constantly being developed and examined by clinical trials.

One of the more common methods of treating cancer, and one that is still used in the majority of cancer cases, is surgery. Surgery may be necessary to stage a tumor, or to remove it from the body. The staging of a tumor using surgery is called a *biopsy*, and involves the removal of a small section of the tumor for examination. Surgery may involve the removal of the entire organ, or just the area surrounding the tumor. In some cases surgery is not an option, because the tumor is either in an inaccessible location or has invaded a critical area of the organ.

Two of the more common forms of cancer treatment are *radiation therapy* and *chemotherapy*. These are frequently used in conjunction with one another. The basic principle of these methods is to attack the cancer cells of the body directly. Chemotherapy is more of a systemic treatment, and is frequently used if the physician is concerned that the tumor has metasticized (begun to spread). In this case, chemicals are used to attack cancer cells wherever they are in the body. These chemicals attack rapidly dividing cells by a variety of mechanisms. Unfortunately, they are usually unable to distinguish between healthy and cancer cells, meaning that the rapidly dividing cells of the hair, blood, and epithelial linings may also be destroyed.

Radiation therapy targets individual tumors in the body, and is frequently used to shrink the size of the tumor prior to surgery or chemotherapy. Radioactive pellets may be placed near the tumor, or radiation is beamed into the intended location. The purpose is to disrupt the cellular processes of the cancer cells, causing them to die and the tumor to shrink.

There are other developing ideas of how to fight cancer. New strategies involve the use of chemicals that interfere specifically with the biochemical pathways of the cancer cells. Other researchers are focusing on using the body's immune system to attack cancer cells, a historically difficult task because the cancer cells are technically part of the body, and therefore not recognized as pathogenic by the immune system. Other research is focusing on using gene therapy to stop cancer cells from dividing. Gene therapy involves introducing genetic material into a cell to replace a defective gene.

In theory, gene therapy may be able to turn off cancer cells by providing a lethal protein or correcting the gene that initially caused the cell to ignore the checkpoints of the cell cycle.

It is likely in the future that cancer will be attacked using a combination of all of these strategies. Most researchers and medical professionals doubt that cancer will ever be cured. Research is instead generally focusing on increasing the five-year survival rate for cancer patients, and then ensuring that the cancer does not return.

Acronyms

ADA American Dental Association

ATP adenosine triphosphate

CAT computerized axial tomography

CCK cholecystokinin

CDC Centers for Disease Control

CF cystic fibrosis

CNS central nervous system

CUP carcinoma of unknown primary

DNA deoxyribonucleic acid

EDTA ethylene diamine tetra acetate

ETC electron transport chain

FDA Food and Drug Administration (U.S.)

GERD gastroesophageal reflux disease

HD Hirschsprung's disease

HDL high-density lipoprotein

GI gastrointestinal

GIP gastric inhibitory peptide

IBS Irritable bowel syndrome

LDL low-density lipoprotein

MMR measles, mumps, and rubella

MRI magnetic resonance imaging

NASA National Aeronautics and Space Administration

NSAID non-steroidal anti-inflammatory drugs

OTC over-the-counter

RNA ribonucleic acid

TNF tumor necrosis factors

USDA U.S. Department of Agriculture

Glossary

Active transport The movement of compounds or ions from an area of low concentration to an area of high concentration across a cell membrane. This process involves the expenditure of energy in the form of adenosine triphosphate (ATP).

Acupuncture A form of medicine, begun in ancient China, which inserts small needles at specific points of the body to treat disease or pain.

Alkaline A term used to indicate a pH of 7 or greater. Sometimes also called *basic*.

Alzheimer's disease A degenerative, non-reversible disease of the nerve cells in the brain. The disease is characterized by a lack of memory, decrease in intellectual abilities, and confusion. There are several forms of the disease, the most common of which effects the elderly and is called late-onset Alzheimer's.

Amino acids The building blocks of proteins. Human proteins are comprised of twenty different types of amino acids, which are linked together by dehydration synthesis reactions to form long chains called peptides.

Amphiphatic A term given to a molecule that has both hydrophilic and hydrophobic properties. An example are the phospholipid molecules of the cell membrane.

Anaerobic A term used to describe reactions that occur in the absence of oxygen.

Antibodies Proteins produced by the immune system that are involved in the response of the body against specific pathogens. These can also be involved in autoimmune diseases against cells of the body.

Antioxidants Compounds that prevent oxidative damage to organic molecules. Vitamins C and E are examples of antioxidant nutrients, as is the mineral selenium.

Asexual A form of reproduction in which a single individual produces offspring. Unlike sexual reproduction, there is no exchange of genetic material with a mate.

Asymptomatic Usually refers to a person who is diagnosed with a disease or ailment, but does not display the symptoms of the disease.

Autism A psychological condition that is characterized by delusions and hallucinations. Autistic people are not able to perceive the environment around them correctly.

Autocatalytic process A chemical reaction in which the products of the reaction are responsible for initiating the start of the reaction.

Autoimmune response A response of the immune system in which the body incorrectly identifies cells or tissues of the body as pathogens and initiates an immune response against them. Also called *autoimmune diseases.*

Barium A metallic element that is frequently used as a diagnostic tool to examine the interior of the gastrointestinal tract. Barium coats the internal tissues, allowing better resolution during an x-ray procedure.

Bilirubin A waste product produced by the liver that is the result of the breakdown of red blood cells. It is released into the small intestine, but some is reabsorbed back into the blood and excreted with the urine. It is the compound responsible for the coloration of the urine and fecal material.

Binucleate A cell that contains two nuclei.

Bioavailability A term of nutritional analysis that indicates how much of a nutrient in a food is actually available to the body for absorption by the gastrointestinal tract.

Biochemistry A branch of science that studies the chemistry of living organisms and the metabolic pathways that are responsible for the processing of nutrients in the body.

Bolus The name given to the mass of food that accumulates at the rear of the oral cavity for swallowing. It proceeds down the esophagus into the stomach, where it becomes chyme.

Buccal cavity Another term commonly used to describe the oral cavity. It technically represents the space between the back of the teeth and gums to the rear of the mouth.

Capillaries The smallest vessels of the circulatory system, and the sites where gas and nutrient exchange is at its greatest. The walls of capillaries are typically one cell layer thick.

Cholesterol A biological molecule that belongs to the class of lipids called the sterols. Cholesterols are important for the formation of many hormones, the synthesis of vitamin D, and the regulation of cellular membranes.

Colonoscopy A medical procedure that is designed to examine the internal structure of the large intestine using a long tube with a camera and light attached to the end. Using this instrument, a physician is able to examine any irregularities in the structure of the large intestine.

Computerized axial tomography (CAT) scan A diagnostic tool in which a radioactive compound is given to a patient and the absorption of the compound by different tissue types is recorded by a detector. It is frequently used in the study of the central nervous system.

Deamination The process by which the amino functional group ($-NH_2$) is removed from a molecule, usually an amino acid or protein.

Deglutition Another term used for the act of swallowing.

Dehydration synthesis A form of chemical reaction that involves the removal of water to form a chemical bond. Also called a *condensation reaction.*

Dentin A tissue that is the majority of the mass of a tooth. It is primarily minerals (70 percent), with the remainder being water and organic material.

Diabetes The common name for diabetes mellitus, a disease that influences carbohydrate metabolism in the body. The disease involves an inadequate production of insulin by the pancreatic cells, and is manifested as an inability to regulate blood glucose levels in the body.

Diaphragm A muscle that aids in respiration. It separates the thoracic cavity from the abdominal cavity.

Diploid Any organism whose cells contain two copies of each chromosome. Usually abbreviated as *2n*, where *n* equals the chromosome number of the species. The majority of human cells, except sex cells and some liver cells, are diploid.

Diuretics Chemical, natural or synthetic, that increase water loss from tissue, which is usually associated with an increase in urine output.

Dominant In genetic analysis, a copy of a gene that masks the expression of a recessive copy of a gene. If the dominant gene is responsible for a disease, usually the person must inherit only a single copy of the gene from one parent in order to display the disease.

Electrolytes Compounds that, when placed in water, form ions, and thus conduct electricity. Electrolytes are important for human metabolism and physiology. Common electrolytes in the blood are potassium and sodium.

Encephalitis An inflammation of the tissues of the brain.

Endocrine The system of hormone-producing glands in the human body. These ductless glands release their chemicals either into the bloodstream, directly onto tissues, or into the lymphatic system.

Endoscope An instrument used by physicians to examine the lumen of the gastrointestinal tract for physical problems, such as ulcers or obstructions. At the end of the endoscope is a small camera and light so that a doctor can visually inspect the surface of the organ.

Enzymes The metabolic assistants to a chemical reaction that are responsible for accelerating the rate of the chemical reaction. Enzymes are usually large proteins with complex shapes that allow them to interact with their target molecules, or substrates.

Epithelial cells A type of cell that lines organs and tissues of the body. It specializes in the exchange of materials with the external environment, such as the lumen of the gastrointestinal tract.

Eructation Also known as burping, this action removes excess gas from the upper regions of the stomach. This gas is usually excess air that enters the stomach during eating.

Exocrine Glands that utilize ducts to release their secretions to the outside environment.

Facilitated diffusion A passive process that utilizes a membrane-bound protein to move a compound across a membrane down its concentration gradient.

Flora In biology, this term is usually used to represent the vegetative organisms occupying a given area. In human biology, it refers to the natural populations of non-pathogenic bacteria that inhabit various areas of the body, including the digestive system.

Gene therapy An advance of biotechnology that involves the delivery of genetic material into a cell for the purpose of either correcting a genetic problem, or giving the cell a new biochemical function.

Genus In the classification of organisms, the group that is just above the level of a species. Humans belong to the genus *Homo*, of which they are the only surviving species, *sapiens*.

Gluconeogenesis The metabolic pathways that generate glucose from nonsugar precursors, such as amino acids.

Glycolysis The first stage of the energy releasing pathways in the cell. Converts glucose into small amounts of energy (ATP) and the molecule pyruvate. Occurs in the cytoplasm of the cell.

Haruspicy The study of the internal organs of animals for the purpose of prophecizing spiritual events.

Hemoglobin The protein found in red blood cells that utilizes iron to transport oxygen to the cells of the body. A single red blood cell may have over 250 million hemoglobin molecules within it.

Hepatic portal system The name given to the portion of the circulatory system that connects the stomach and both intestines to the liver.

Hermaphroditic Containing both male and female reproductive organs. Hermaphroditic organisms rarely mate with themselves.

Histamine A chemical signal of the immune system that serves to increase the fluid content of tissues by allowing more fluids to leave the circulatory system.

Homeostasis A property of metabolism by which an organism maintains the internal environment within certain parameters despite fluctuations in the external environment. A good example in humans is body temperature or blood glucose levels.

Hormones A chemical that is secreted by one gland of the body to influence the operation of a second, usually more distant, gland or organ. Hormones are secreted by endocrine glands into the bloodstream.

Human Genome Project A groundbreaking project to construct a genetic map of the entire human genome. The project was announced completed in 2003, although work on various aspects of the project will continue for years.

Hydrolysis In chemistry, the breaking of a chemical bond by the addition of water.

Hydrophilic A water-loving compound, meaning that it is soluble in water. An example is glucose.

Hydrophobic A water-fearing compound, meaning that is it generally insoluble in water. Most lipids are hydrophobic, as are some amino acids.

Hypertension A disease that is characterized by chronically elevated blood pressure. It is frequently associated with diet, although there are other risk factors for the disease.

Intrinsic factor The term given to the factor produced by the stomach to assist in vitamin B_{12} absorption. Without this factor, very little of the vitamin is absorbed, leading to pernicious anemia.

Invertase An enzyme discovered in the nineteenth century that breaks down the sugar sucrose into glucose and fructose. This was the first enzyme to be discovered inside a cell that also had activity outside the cell.

Ions Any element or compound that loses or gains electrons and in the process changes its net electric charge.

Isomers Chemical compounds that have the same composition of elements but different structures. The monosaccharides are all isomers of one another because they share the same chemical formula ($C_6H_{12}O_6$) but have different structures.

Kilocalorie The amount of energy required to raise 1,000 grams of water from 14.5° to 15.5° Celsius at standard atmospheric pressure. The unit of energy measurement that is commonly used in nutritional analysis.

Lacteals The portion of the lymphatic system that is associated with the gastrointestinal system, specifically the intestines.

Laser A very narrow, intense beam of light that is either in the infrared or visible wavelengths of the electromagnetic spectrum.

Ligament The dense connective tissue that connects muscles to bone.

Lipoproteins Proteins that are connected chemically to lipids. Lipoproteins are used by the digestive system to transport hydrophobic fats and lipids in the hydrophilic blood stream.

Lumen The name given to the internal space within the gastrointestinal tract. The lumen is actually an external environment, although its conditions (temperature and pH) are carefully regulated by the body. The lumen is the site of digestion, with nutrients then being absorbed into the tissues of the intestines.

Lymphocytes The name given to those cells that are involved in the immune response. Lymphocytes are involved in the defense against specific pathogens. Both B and T cells are lymphocytes.

Lymphoma The name given to a cancer of the lymphatic system. Lymphoma is actually a cancer of both the lymphatic and circulatory systems, since it can affect cells common to both systems.

Magnetic resonance imaging (MRI) A diagnostic tool based on the fact that hydrogen atoms resonate or vibrate at distinct frequencies when bombarded by energy from a high power magnet. This tool produces a three-dimensional image of the tissues being studied, and is very useful in examining minor variances in body chemistry.

Manometer A device that measures the difference in pressure between two fluids.

Medulla oblongata A portion of the brain stem that is involved in the involuntary processes of respiration, circulation, and digestion.

Meningitis An inflammation of the meninges, a series of membranes that envelope and protect the central nervous system. Meningitis may be caused by bacteria, viruses, or protozoa.

Mesentery A tissue that suspends the digestive glands within the abdominal cavity. The mesentery connects to the outer layer of the gastrointestinal tract.

Metabolism The sum of all of the chemical reactions in a cell, tissue, organ, or organism. In nutritional terms, it frequently applies to the processing of the energy nutrients and generation of energy.

Metastasis The movement of cancer cells from a primary tumor to another tissue or organ of the body, forming a secondary tumor.

Mitochondria An organelle of the cell that is responsible for the majority of the energy-releasing processes. This organelle generates large amounts of ATP by the process of aerobic respiration.

Multiple sclerosis A degenerative disease of the nervous system in which the myelin covering of neurons deteriorates.

Mumps A viral disease that infects the salivary glands. In most instances it infects the parotid salivary glands located just in front of the ears and above the jaws.

Organelle A membrane-bound compartment within a cell. Organelles are the site of specialized chemical reactions that may be highly regulated by the cell.

Organic In chemistry, compounds that contain carbon-carbon or carbon-hydrogen bonds. In nutritional studies, this term frequently indicates a food that is grown in the absence of chemicals, including pesticides.

Organic chemistry The study of chemical reactions involving carbon compounds. The study of organic chemistry that relates to living organisms is frequently called biochemistry.

Osmosis The diffusion of water from an area of high concentration to low. The movement of water typically is the inverse of a gradient of salt or sugar.

Oxidation A chemical reaction that involves the loss of electrons. The oxidation of one compound is usually is coupled to the reduction (electron gain) of another compound.

Papillae The raised areas of the tongue. Depending on the region of the tongue, the base of the papillae house the taste buds.

Pasteurization The use of heat to remove the microbial content of a food or liquid. Milk is commonly pasteurized.

Pedigree In genetic analysis, the tracing of the ancestral lineage of an individual. The process often utilizes a graphic representation, also called a family tree.

Peristaltic action A rhythmic contraction of the muscles of the gastrointestinal tract, most notably in the small intestine, that is responsible for moving nutrients and undigested material through the lumen towards the anus.

pH The acidity of a solution. It is formally the measure of the hydrogen ion concentration of a solution, indicated by the formula: $pH = -\log [Hv]$.

Phagocytosis A mechanism by which a cell can engulf material. A pocket forms in the cell membrane and gradually moves within the cytoplasm forming a vesicle. The contents of the vesicle may then be digested or processed by cellular enzymes.

Pharynx The rear area of the oral cavity. This area connects the respiratory and digestive systems of the body.

Phospholipids A class of organic molecules that resemble triglycerides but have one fatty acid chain replaced by a phosphate group. The structure of the phospholipid gives the molecule both hydrophobic and hydrophilic regions.

Phylum In the classification of organisms, a phylum represents a major group, just under the level of a kingdom. Organisms are placed into phylums based upon very general characteristics. Frequently, these characteristics represent major evolutionary events, such as the formation of a body cavity.

Polyploid In humans, each cell has two copies of each chromosome, one maternal and one paternal. Polyploid indicates multiples of chromosomes more than two in a cell.

Pons An area of the brain stem that serves as a bridge between the spinal column and the higher areas of the brain.

Protistans A kingdom of life that is characterized as generally being single-celled organisms with a nucleus. Members of this kingdom are only distantly related to one another. Some protistans are parasites of the digestive tract.

Proto-enzyme This term usually refers to a protein that is not yet activated to form an enzyme. Many of the digestive enzymes formed by the pancreas are proto-enzymes, which are activated once in the lumen of the small intestine. The delay in activation protects the delicate tissues of the pancreas from damage.

Protozoa The name that is commonly used for animal-like protistans. These are single-celled organisms that possess a nucleus. Some of these are parasites of humans.

Radiologist A medical professional that specializes in the use of electromagnetic radiation, including x-rays, for the diagnosis and treatment of diseases.

Recessive In genetic analysis, a copy of a gene that is masked by a dominant copy of a gene. If the recessive gene is responsible for a disease, usually the person must inherit the same recessive gene from both parents in order to display the disease.

Rickets A deficiency disease of children that is caused by inadequate vitamin D in the diet. This deficiency causes a decrease in calcium absorption by the small intestine, resulting in improper bone formation.

Saturation In chemistry, this term is frequently used to describe the structure of fatty acid chains in triglycerides. A saturated fatty acid does not possess any double bonds and contains the maximum number of hydrogen atoms per carbon. A monounsaturated contains one double bond in the fatty acid, while a polyunsaturated has multiple double bonds. The level of saturation influences the metabolic characteristics of the fatty acid.

Scrotum In males, the sac of tissue that holds the reproductive structures, including the testes.

Scurvy A disease caused by a deficiency in vitamin C (ascorbic acid) that results in breakdown of connective tissue such as collagen, gingivitis, anemia, and skin problems.

Skeletal muscle The type of muscle in the human body that is under voluntary control.

Spore A part of the life history of some microbes that is highly resistant to damage from the environment. It is characteristic of a group of organisms called the Sporozoans, common parasites of the intestinal tract.

Stroke This term refers to any problem with the circulatory system in the brain that inhibits the flow of blood to the tissues of the brain.

Symbiotic A relationship between two unrelated species of organisms. The relationship may be either beneficial for both species, or harmful for one of the species.

Sympathetic nervous system The organizational section of the nervous system that is responsible for the fight-or-flight response. The sympathetic nervous system controls smooth muscle, which is not under voluntary control of the body.

Transposon A mobile genetic element. These small pieces of genetic material may move around the genome, disrupting the function of genes. Some studies suggest that they may be involved in some forms of cancer.

Ultrasound A medical procedure by which sound waves are applied to the body. The reflections of these waves are detected by a specialized instrument and provide a view of the internal structures of the body. This procedure is useful in examining tissues that normally would not be detectable by x-ray examination.

Vagus nerve The tenth (X) cranial nerve that is responsible for smooth muscle contraction and the control of digestive secretions. It is also involved in the process of swallowing.

Velocardiofacial syndrome A type of birth defect that has a wide range of effects on the body. The most common problems, are a cleft palate or lip, heart problems and altered facial features.

Vestibular apparatus The portion of the inner ear that is responsible for detecting the orientation of the body in space. The vestibular apparatus also detects acceleration.

Vestigial In human biology, an organ that no longer appears to have a functional purpose. It represents the remains of an organ that had some purpose in the evolutionary history of the species.

Villi The protrusions of the mucosa layer of the intestines (primarily the small) that serve as the site of nutrient absorption. Each villi contains components of the circulatory and lymphatic system, as well as many smaller cellular projections called microvilli.

X-ray A form of electromagnetic radiation that is frequently used by radiologists to examine the internal structures of the body. It is very useful in detecting bone and connective tissue, but is limited in the examination of softer tissues.

Zoology A branch of the biological sciences that examines the physiology, anatomy, genetics, biochemistry, and relationships among members of the animal kingdom.

Zymase An enzyme involved in the fermentation of sugars into alcohols.

Organizations and Web Sites

American Academy of Periodontology
737 N. Michigan Ave., Suite 800
Chicago, IL 60611-2690
Phone: (312) 787-5518
Fax: (312) 787-3670
Email: member.services@perio.org
www.perio.org

This organization's site contains a wealth of information on periodontal disease, from symptoms and treatments to research studies for professionals.

American Cancer Society
1810 Woodfield Dr., Suite 100
Savoy, IL 618745
Phone: (800) ACS-2345
www.cancer.org

A comprehensive site for all forms of cancer. Includes information on types of cancer, including statistics and risk factors. Treatments and mechanisms of getting connected with cancer survivors are also provided.

American Dental Association
211 East Chicago Ave.
Chicago, IL 60611-2678
Phone: (312) 440-2500
www.ada.org

The most interesting section of this site is located under the ePublic links and includes descriptions of most ailments and diseases of the oral cavity.

Bad Bug Book
www.wikipedia.org/wiki/Bad_Bug_Bookvm.cfsan.fda.gov/~mow/intro.html

A number of these sites exist on the Internet. Each contains information on foodborne pathogens, symptoms, and preventative measures. Most are written for the general audience and some contain links to more scientific papers.

California Hepatitis Resource Center
Email: info@hepatitisresources-calif.org
www.hepatitisresources-calif.org

An excellent online resource for individuals who want to learn more about the various forms of hepatitis. Provides demographic information on the rates of various forms in California.

Celiac Disease Foundation
13251 Ventura Blvd., #1
Studio City, CA 91604
Phone: (818) 990-2354
Fax: (818) 990-2379
Email: cdf@celiac.org
www.celiac.org

An organization dedicated to providing information on celiac disease. Contains links to online copies of quarterly newsletters. Contains information on ongoing research programs.

Cleft Palate Foundation
104 South Estes Dr., Suite 204
Chapel Hill, NC 27514
Phone: (800) 24-CLEFT or (919) 933-9044
Email: info@cleftline.org
www.cleftline.org

Provides a series of online publications for parents and families of children with cleft palate, as well as links to other organizations that assist families and individuals.

Crohn's and Colitis Foundation of America
386 Park Ave. South, 17th Floor
New York, NY 10016
Phone: (800) 932-2423
Email: info@ccfa.org
www.ccfa.org

A gateway to information on these diseases. Visitors must register for access, but the site provides useful information on treatments for the disease, as well as links to professionals and local chapters for support groups.

GERD Information Resource Center
www.gerd.com

A site maintained by AstraZeneca LP that contains information on gastroesophageal reflux disease (GERD), as well as information for health care providers.

Intestinal Disease Foundation
100 West Station Square Dr.
Pittsburgh, PA 15219-1122
Phone: (412) 261-5888 or (877) 587-9606
Fax: (412) 471-2722
Email: info@intestinalfoundation.org
www.intestinalfoundation.org/library/diverticular.shtml

This link provides access to diverticular disease, although the foundation is involved with all aspects of inflammatory bowel diseases. Contains both a bookstore and on-line library.

Irritable Bowel Syndrome Self Help Group
www.ibsgroup.org

Online chat groups, discussion forms, bookstores and information guides regarding IBS. Designed to assist individuals with IBS get answers to their questions and concerns. Also contains links to recent articles on IBS.

National Cancer Institute
NCI Public Inquiries Office
Suite 3036A
6116 Executive Blvd., MSC8322
Bethesda, MD 20892-8322
Phone: (800) 422-6237
cancer.gov

A similar site to the American Cancer Society, this site serves as an information clearinghouse for all aspects of cancer, from symptoms to treatment.

National Institute of Diabetes & Digestive & Kidney Diseases (NIDDK)
Office of Communications and Public Liaison
NIDDK, NIH
Building 31, Room 9A04
Center Drive MSC 2560
Bethesda, MD 20892-2560
Phone: (301) 496-3583
www.niddk.nih.gov

A site managed by the National Institutes of Health (NIH) that contains useful information on a wide variety of common and rare digestive disorders, including ulcers and *H. pylori* infections and treatments. Entries are targeted at the general audience.

Nobel e-Museum
www.nobel.se

A searchable list of all Nobel Prize winners as well as information about the Nobel Prize process.

Bibliography

Asimov, Isaac. *Asimov's Chronology of Science and Discovery*. New York: HarperCollins, 1994.
———. *Asimov's New Guide to Science*. New York: Basic Books, 1984.
Bellenir, Karen. *Digestive Diseases and Disorders Sourcebook: Basic Consumer Health Information about Diseases that Impact the Upper and Lower Digestive System*. Detroit, MI: Omnigraphics, 2000.
Blaser, Martin J. "The Bacteria Behind Ulcers." *Scientific American* 274, no. 2 (February 1996): 104–109.
Bynum, W. F., E. J. Browne and R. Porter. *Dictionary of the History of Science*. Princeton, NJ: Princeton University Press, 1981.
Castiglioni, Arturo. *A History of Medicine*. New York: Alfred A. Knopf, 1946.
Christensen, Damaris. "Is Your Stomach Bugging You?" *Science News* 156 (October 1999): 234–236.
Cook, Alan R., ed. *The New Cancer Sourcebook*. Detroit, MI: Omnigraphics, 1996.
———. *Oral Health Sourcebook: Basic Information about Diseases and Conditions Affecting Oral Health*. Detroit, MI: Omnigraphics, 1998.
Edmundowciz, Steven. *20 Common Problems in Gastroenterology*. New York: McGraw-Hill, 2002.
Gillespie, Stephen H., and Richard D. Pearson, eds. *Principles and Practice of Clinical Parasitology*. New York: John Wiley & Sons, 2001.
Heelan, Judith S., and Frances W. Ingersoll. *Essentials of Human Parasitology*. Albany, NY: Delmar, 2002.
Johanson, John F, ed. *Gastrointestinal Diseases: Risk Factors and Prevention*. Philadelphia, PA: Lippincott-Raven, 1997.
Katz, Michael, Dickson D. Despommier, and Robert Gwadz. *Parasitic Diseases*, 2nd ed. New York: Springer-Verlag, 1989.
Knight, Bernard. *Discovering the Human Body*. New York: Lippincott & Crowell, 1980.
Kondo, Shinji, Brian C. Shutte, Rebecca J. Richardson, et al. "Mutations in IFR6

cause Van der Woude and popliteal pterygium syndromes." *Nature Genetics* 32, no. 2 (2002): 285–289

Lane, Samantha, and David Lloyd. "Current Trends in Research into the Waterborne Parasite *Giardia*." *Critical Reviews in Microbiology* 28, no. 2 (2002): 123–147.

McKinnell, Robert, Ralph Parchment, Alan Perantoni, and G. Barry Pierce. *The Biological Basis of Cancer*. Cambridge, UK: Cambridge University Press, 1998.

Murphy, Gerald, Lois Morris, and Dianne Lange. *Informed Decisions: The Complete Book of Cancer Diagnosis, Treatment and Recovery*. New York: Viking, 1997.

Payne, W. S., and A. M. Olsen. *The Esophagus*. Philadelphia, PA: Lea & Febiger, 1974.

Porter, Ray. *The Greatest Gift to Mankind: A Medical History of Humanity*. New York: W.W. Norton & Company, 1997.

Raven, Peter H., and George B. Johnson. *Biology*, 5th ed. Boston, MA: WCB/McGraw-Hill, 1999.

Ronan, Colin A. *Science: Its History and Development Among the World's Cultures*. New York: Facts on File, 1982.

Sherwood, Lauralee. *Human Physiology: From Cells to Systems*, 4th ed. Pacific Grove, CA: Brooks/Cole, 2001.

Singer, Charles. *A Short History of Anatomy from the Greeks to Harvey*. New York: Dover Publications, 1957.

Symons, Alan. *Nobel Laureates, 1901–2000*. London: Polo, 2000.

Taton, Rene, ed. *History of Science: Ancient and Medieval Science from the Beginnings to 1450*. New York: Basic Books, 1964.

———. *History of Science: The Beginnings of Modern Science from 1450 to 1800*. New York: Basic Books, 1964.

———. *History of Science: Science in the Nineteenth Century*. New York: Basic Books, 1964.

———. *History of Science: Science in the Twentieth Century*. New York: Basic Books, 1964.

Tortora, Gerald J., and Sandra R. Grabowski. *Principles of Anatomy and Physiology*, 8th ed. New York: HarperCollins, 1996.

Whitney, Eleanor J., and Sharon R. Rolfes. *Understanding Nutrition*, 8th ed. Belmont, CA: Wadsworth, 1999.

Windelspecht, Michael. *Groundbreaking Scientific Experiments, Inventions and Discoveries of the 19th Century*. Westport, CT: Greenwood, 2003.

———. *Groundbreaking Scientific Experiments, Inventions and Discoveries of the 17th Century*. Westport, CT: Greenwood, 2002.

Wolf, A. *A History of Science, Technology, and Philosophy in the 18th Century*. Gloucester, MA: Peter Smith, 1968.

———. *A History of Science, Technology, and Philosophy in the 16th & 17th Centuries*. Gloucester, MA: Peter Smith, 1968.

Index

About the Author

MICHAEL WINDELSPECHT is Assistant Professor of Biology at Appalachian State University. He is the author of two books in Greenwood's *Groundbreaking Scientific Experiments, Inventions, and Discoveries through the Ages* series, coauthor of *The Lymphatic System* in the *Human Body Systems* series, and editor of the *Human Body Systems* series.